To our families, teachers, and fellows

D.T.A.
R.F.S.
W.J.S.

Excimer Laser Phototherapeutic Keratectomy

Dimitri T. Azar, M.D.
Director of Corneal and Refractive Surgery Services
and Associate Chief of Ophthalmology
Massachusetts Eye and Ear Infirmary
Associate Professor of Ophthalmology
Harvard Medical School
Boston, Massachusetts

Roger F. Steinert, M.D.
Surgeon
Boston Eye Surgery and Laser Center
Ophthalmic Consultants of Boston
Associate Surgeon
Massachusetts Eye and Ear Infirmary
Assistant Clinical Professor
Harvard Medical School
Boston, Massachusetts

Walter J. Stark, M.D.
Director of Corneal Service
Wilmer Ophthalmological Institute
Johns Hopkins Hospital
Professor of Ophthalmology
Johns Hopkins University
Baltimore, Maryland

Editor: Darlene Barela Cooke
Managing Editor: Frances M. Klass
Marketing Manager: Diane M. Harnish
Production Coordinator: Dana M. Soares
Typesetter: Better Graphics, Inc.
Printer: Quebecor Printing

Copyright © 1997 Williams & Wilkins

351 West Camden Street
Baltimore, Maryland 21201-2436 USA

Rose Tree Corporate Center
1400 North Providence Road
Building II, Suite 5025
Media, Pennsylvania 19063-2043 USA

All rights reserved. This book is protected by copyright. No part of this book may be reproduced in any form or by any means, including photocopying, or utilized by any information storage and retrieval system without written permission from the copyright owner.

Accurate indications, adverse reactions and dosage schedules for drugs are provied in this book, but it is possible that they may change. The reader is urged to review the package information data of the manufacturers of the medications mentioned.

Printed in United States

First Edition

Library of Congress Cataloging-in-Publication Data

Azar, Dimitri T.
    Excimer laser phototherapeutic keratectomy / Dimitri T. Azar,
Roger F. Steinert, Walter J. Stark.
        p.   cm.
    Includes bibliographical references and index.
    1. Eye—Laser surgery. 2. Eximer lasers. I. Steinert, Roger F.
II. Stark, Walter J.   III. Title.
    [DNLM: 1. Keratectomy, Photorefractive, Phototherapeutic Excimer Laser—methods.
WW 220 A992e 1997]
RE86.A93   1997
617.7'19—dc21
DNLM/DLC
for Library of Congress   96-40065
                                                                                     CIP

The publishers have made every effort to trace the copyright holders for borrowed material. If they have inadvertently overlooked any, they will be pleased to make the necessary arrangements at the first opportunity.

To purchase additional copies of this book, call our customer service department at **(800) 638-0672** or fax orders to **(800) 447-8438**. For other book services, including chapter reprints and large quantity sales, ask for the Special Sales department.

Canadian customers should call **(800) 665-1148**, or fax **(800) 665-0103**. For all other calls originating outside of the United States, please call **(410) 528-4223** or fax us at **(410) 528-8550**.

Visit Williams & Wilkins on the Internet: http://www.wwilkins.com or contact our customer service department at custserv@wwilkins.com. Williams & Wilkins customer service representatives are available from 8:30 am to 6:00 pm, EST, Monday through Friday, for telephone access.

97 98 99 00
1 2 3 4 5 6 7 8 9 10

Foreword

In ophthalmology innovation has not always been rapid. Concepts, treatments, and technical capacities have evolved slowly over decades or centuries. It is therefore amazing how rapidly laser technology has been applied as a treatment modality in our specialty, and, particularly, how easily excimer photorefractive keratectomy has been accepted into mainstream ophthalmology. The next step in the application of this laser to removal of corneal surface pathologies (opacities, irregularities)—phototherapeutic keratectomy (PTK)—was a given and this is why the present text is so important and timely.

Whether PTK will in one sweep replace other modalities which have served humanity for a long time is not entirely certain. Superficial keratectomy in cases of stromal opacities close to the surface has a long and interesting history—it was advocated as early as Galen (abrasio), and, as late as the 18th century, grinding with powdered glass was still recommended. Later, the removal of stromal surface layers by surgery was introduced by von Walther in 1840 and others, and it can now certainly be done with great precision. If the deepest stroma also has to be excised, a lamellar transplant offers advantages in preserving corneal thickness and restoring surface smoothness. This development was pioneered particularly by von Hippel (1888) and in this century by Paufique and other French surgeons. Although lamellar keratoplasty is a safe procedure, since the graft is rarely vulnerable to immune reaction, it is technically demanding and time consuming. In addition, the resulting visual acuity is often less than after penetrating keratoplasty in a similar situation. The latter technique has been a surgical success story since Zirm presented the first (documented) clear corneal graft in 1906 and is now usually the method of choice in corneal opacity. Still penetrating keratoplasty can be hazardous and, therefore, for superficial lesions, the advantages of a simpler technique like PTK are numerous.

Irregular astigmatism has traditionally been dealt with by applying hard contact lenses. This development began in the first half of this century with large scleral lenses made of glass. They were heavy and difficult to fit and they were replaced in the 1930s by plastic material. A decade later, Tuohy introduced corneal lenses as we know them today. Later, gas permeable materials increased tolerance. However, contact lenses on irregular corneas are not always tolerated, and even when they are, they are often perceived by the patient as a hassle and expense. Will a quick shave with the excimer laser solve the problem? Read on for the answer.

We are indebted to Dr. Dimitri Azar, who is an authority on refractive and therapeutic corneal surgery, and to Drs. Roger Steinert and Walter Stark, who are leaders in corneal and anterior segment surgery. They have contributed to many chapters, have expertly edited this timely text on PTK, and have involved expert co-authors. This text will be a classic in the field for a long time.

Claes H. Dohlman, M.D.

Preface

This book describes the principles and practice of *Excimer Laser Phototherapeutic Keratectomy* (PTK) for the treatment of corneal scars, dystrophies, surface irregularities and PRK complications. The clinical aspects of PTK are emphasized, including preoperative findings, indications, surgical techniques, clinical results and complications.

The use of the ArF excimer laser has revolutionized the fields of refractive and corneal surgery. It was introduced in 1983 as a means of removing or ablating corneal tissue with excellent precision. Subsequent work showed that this is accompanied by minimal injury to the adjacent tissue. Investigations of excimer laser phototherepeutic keratomy (PTK) in humans were sponsored, and subsequently approved, by the U.S. Food and Drug Administration. It became clear that hyperoptic shifts and other complications may occur, but if performed properly, the side effects of PTK can be minimized and the need for subsequent penetrating keratoplasty can be reduced. This textbook will not only describe the complications and refractive alterations of PTK, but will also describe the use of PTK in the management of optical and refractive alterations caused by other refractive surgical procedures.

The introductory section of the book presents an overview of the history and development of excimer lasers, as well as a summary of corneal wound healing and light scattering. Detailed in the sections to follow are indications and techniques of phototherapeutic keractectomy, including discussions and illustrations of the treatment of corneal irregularities, nodules and scars. A review of PRK outcomes and complications is followed by discussion of the indications and techniques of managing PRK complications including central islands, undercorrection, sub-epithelial haze and treatment decentrations.

We would like to express our gratitude to the contributors who were willing to spend long hours researching, writing, editing, and re-editing. We believe that in so doing, they have broadened the perspective and utility of their chapters, and improved the quality of the book.

We would like to thank our managing editor, Phyllis Friello, for her efforts in reviewing the manuscripts and in communicating with the publisher and the contributors. We would also like to acknowledge the efforts of Beth Barry and Igaku Shoin in making this project possible. They have managed to keep us and other contributors on schedule.

We have included numerous pre-operative and post-operative photographs to illustrate the principles and techniques of surgery providing the PTK surgeon with a reference for treatment indications and outcomes. It is our hope that this book will help facilitate the learning curve of PTK and bridge the gap between the art and science of this extremely useful technology.

Dimitri Azar, M.D.
Roger Steinert, M.D.
Walter Stark, M.D.

Contributors

M. Farooq Ashraf, M.D.
Department of Ophthalmology
Johns Hopkins University
Wilmer Eye Institute
Baltimore, Maryland

Dimitri T. Azar, M.D.
Director of Corneal and Refractive Surgery
  Services
Massachusetts Eye and Ear Infirmary
Associate Professor of Ophthalmology
Harvard Medical School
Boston, Massachusetts

Perry S. Binder, M.D.
Associate Clinical Professor
Department of Ophthalmology
University of California, San Diego and the
  National Vision Research Institute
San Diego, California

Tat Keong Chan, F.R.C.S., F.R.C.Ophth.
Senior Lecturer
Department of Ophthalmology
National University of Singapore
Singapore, Singapore

Shu-Wen Chang, M.D.
Research Fellow
Department of Ophthalmology
Wilmer Eye Institute
Johns Hopkins University
Baltimore, Maryland

Ezra L. Galler, M.D.
Director of Corneal and Refractive
  Surgery
Koch Eye Associates
Warwick, Rhode Island

Marco C. Helena, M.D.
Cornea Fellow
Massachusetts Eye and Ear Infirmary
Department of Ophthalmology
Harvard Medical School
Boston, Massachusetts

Sandeep Jain, M.D.
Department of Ophthalmology
Columbia Presberterian Medical Center
The Edward S. Harkness Eye Institute
New York, New York

Ernest W. Kornmehl, M.D.
Associate Clinical Professor
Department of Ophthalmology
Tufts University School of Medicine
Co-Director, Novatec Laser Program
Massachusetts Eye and Ear Infirmary
Boston, Massachusetts

Andrew Martin, M.D.
Greater Baltimore Medical Center
Department of Ophthalmology
Baltimore, Maryland

Russell McCally, Ph.D.
Associate Professor
Department of Ophthalmology
Wilmer Eye Institute
Principal Staff Physicist
Applied Physics Laboratory
Johns Hopkins University
Baltimore, Maryland

Marc Odrich, M.D.
Medical Monitor
VISX Incorporated
Santa Clara, California
Associate Professor of Ophthalmology
Columbia Presberterian Medical Center
The Edward S. Harkness Eye Institute
New York, New York

Christopher J. Rapuano, M.D.
Associate Professor
Jefferson Medical College of
 Thomas Jefferson University
Associate Surgeon
Cornea Service, Wills Eye Hospital
Philadelphia, Pennsylvania

Michael Rogers, M.S.
Department of Biomedical Engineering and
 Applied Physics Laboratory
Johns Hopkins University
Laurel, Maryland

Riad Shalash, M.D.
Research Fellow
Department of Ophthalmology
Wilmer Eye Institute
Johns Hopkins University School of Medicine
Baltimore, Maryland

Walter Stark, M.D.
Director of Corneal Service
The Wilmer Ophthalmological Institute
Johns Hopkins Hospital
Professor of Ophthalmology
Johns Hopkins University
Baltimore, Maryland

Roger F. Steinert, M.D.
Surgeon
Boston Eye Surgery and Laser Center
Ophthalmic Consultants of Boston
Associate Surgeon
Massachusetts Eye and Ear Infirmary
Assistant Clinical Professor
Harvard Medical School
Boston, Massachusetts

Mary Ann Stepp, Ph.D.
Associate Professor of Ophthalmology
 and Biochemistry
George Washington University
Washington, D.C.

Jonathan H. Talamo, M.D.
Cornea Consultants
Massachusetts Eye and Ear Infirmary
Assistant Clinical Professor
Harvard Medical School
Boston, Massachusetts

Suhas W. Tuli, M.D.
Research Fellow
Department of Ophthalmology
Wilmer Eye Institute
Johns Hopkins University School of Medicine
Baltimore, Maryland

Joshua A. Young, M.D.
Massachusetts Eye and Ear Infirmary
Department of Ophthalmology
Harvard Medical School
Boston, Massachusetts

Contents

Section I.   Introduction

    1.   Excimer Laser Optics and Corneal Applications   3
Riad Shalash, M.D.
Andrew Martin, M.D.
Dimitri T. Azar, M.D.

    2.   Corneal Anatomy, Transparency, and Light Scattering   21
Ezra L. Galler, M.D.
Roger F. Steinert, M.D.

    3.   Corneal Wound Healing After Excimer Keratectomy   33
M. Farooq Ashraf, M.D.
Sandeep Jain, M.D.
Mary Ann Stepp, Ph.D.
Dimitri T. Azar, M.D.

Section II.   Preoperative Evaluation

    4.   Corneal Topography in Phototherapeutic Keratectomy   51
Michael Rogers, M.S.
Russell McCally, Ph.D.
Dimitri T. Azar, M.D.

    5.   Preoperative and Postoperative Protocols   65
Christopher J. Rapuano, M.D.

Section III.   PTK for Corneal Disorders

    6.   Anterior Corneal Dystrophies: Clinical Features   73
Suhas Tuli, M.D.
Shu-Wen Chang, M.D.
Walter J. Stark, M.D.
Dimitri T. Azar, M.D.

    7.   Anterior Corneal Dystrophies: PTK Techniques   99
Sandeep Jain, M.D.
Dimitri T. Azar, M.D.
Walter J. Stark, M.D.

8. **Corneal Nodules and Scars** 117
Joshua A. Young, M.D.
Ernest W. Kornmehl, M.D.

9. **Recurrent Erosion Syndrome** 133
Suhas W. Tuli, M.D.
Dimitri T. Azar, M.D.
Walter J. Stark, M.D.
Perry S. Binder, M.D.

10. **PTK Complications** 143
Marco C. Helena, M.D.
Jonathan H. Talamo, M.D.

Section IV. PTK for PRK Complications

11. **Photorefractive Keratectomy (PRK) Outcomes and Complications** 157
Tat Keong Chan, F.R.C.S., F.R.C.Ophth.
M. Farooq Ashraf, M.D.
Dimitri T. Azar, M.D.

12. **PTK in the Management of PRK Complications** 175
Dimitri T. Azar, M.D.
Walter J. Stark, M.D.
Roger F. Steinert, M.D.

Section V. FDA-Sponsored PTK Clinical Trials

13. **PTK Results: Summit Excimer Laser** 191
Roger F. Steinert, M.D.

14. **PTK Results: VISX Excimer Laser** 201
M. Farooq Ashraf, M.D.
Dimitri T. Azar, M.D.
Marc Odrich, M.D.

section one
Introduction

CHAPTER 1

Excimer Laser Optics and Corneal Applications

Riad Shalash, Andrew Martin, Dimitri T. Azar

INTRODUCTION

Corneal laser surgery remains at the forefront of innovations in ophthalmology and continues to generate interest in investigative and clinical activities. This interest has its early origins in 1973, when Cherry et al.[1] reported the use of the argon laser to treat corneal neovascularization. Since that time, new lasers have been developed and investigated for therapeutic and refractive corneal surgery.

In 1975, Velasco and Sester[2] noted that certain physical properties of the meta-stable states of noble or rare gas atoms resembled those of the alkali metals. They reported on the similarity of the chemical properties of the meta-stable xenon (Xe) atoms to those of sodium and lithium in their ability to react with halogens (such as fluorine) to produce unstable compounds, Xe-halides. These compounds would dissociate rapidly to the ground state and release energetic ultraviolet photons. Velasco and Sester concluded that the diatomic noble gas–halides were of special interest because these bound-free emissions have considerable potential as ultraviolet laser systems.

A few months after they suggested that these meta-stable compounds could be used as lasers, four molecules—xenon fluoride (XeF), xenon chloride (XeCl), xenon bromide (XeBr), and krypton fluoride (KrF)—were observed to undergo light amplification by stimulated emission (i.e., lasing) when excited by an electron beam under appropriate conditions.[3–5] Laser action at 1933 Å (193.3 nm) from the argon fluoride molecule was reported later by Hoffman et al.[6]

The term *excimer* had been coined by Stevens and Hutton[7] in 1960 (short for *excited dimer*) to describe an energized molecule with two identical components. The term *excimer* became the common name for the family of lasers based on this class of compounds. This name was applied to noble gas–halide lasers and persisted even though it is a misnomer. (The lasing medium is a combination of two different elements, a noble gas and a halide, rather than a dimer.) More accurate but less popular names have been used for these lasers, including *rare gas halide lasers*, which describes the gas mixture in the cavity, and the name of the specific gas mixture (e.g., *argon-fluorine laser*) to describe a specific mixture.

The energy used to operate the early excimer lasers was a 2-MeV electron beam with 6 kJ of energy which was provided by a linear accelerator. This system was of limited use for biological research. Burnham and Djeu[8] used transverse electrical discharge to demonstrate laser action with ArF in a relatively compact device, dramatically reducing the required housing space. By 1978, both Lambda Physik (Göttingen, Germany) and Tachisto were marketing excimer lasers for laboratory use.

HISTORICAL REVIEW OF CORNEAL LASER APPLICATIONS

Although the cellular changes induced by light energy may result in harmful effects (such as skin cancer), these effects may be of great physiological and therapeutic value. We now have available for such beneficial use laser systems that may alter corneal tissue using photothermal, photochemical, or photodisruptive mechanisms (Figure 1.1).

Corneal Photocoagulation

Photocoagulation occurs as the light energy is absorbed by a target tissue and converted to heat.

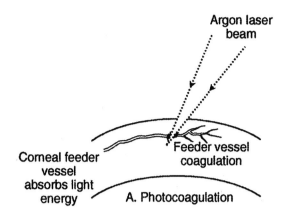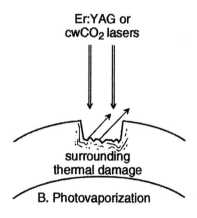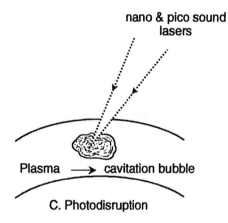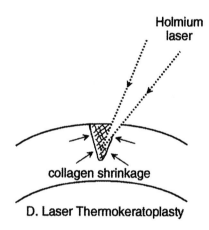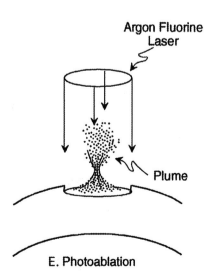

Figure 1.1. Methods of laser–cornea interactions. *A:* Photocoagulation. *B:* Photovaporization. *C:* Photodisruption. *D:* Laser thermokeratoplasty. *E:* Photoablation.

Ophthalmologists are familiar with the effects produced by the argon laser in the treatment of retinopathy and in laser trabeculoplasty. The reactions range from protein denaturization to carbonization. Newer lasers such as solid-state diode lasers and dye lasers produce tissue effects through photocoagulation.

Since the cornea is transparent to visible light, photocoagulation has minimal applications in corneal surgery. Several investigators reported results of corneal laser photocoagulation following Cherry's report of the use of the argon laser to treat corneal neovascularization: Marsh and Marshall[9] reported the treatment of the feeder vessels in patients with lipid keratopathy; and Bahn et al.[10] reported the use of argon laser photocoagulation to lyse vitreous adhesions to corneal wounds after trauma. In addition, Baer and Foster[11] reported success in using a yellow dye laser at 577 nm for photocoagulation of corneal neovascularization (Figure 1.1.A).

Corneal Photovaporization

Longer infrared laser light can produce another type of laser tissue interaction, photovaporization, which allows excision of corneal collagen lamellae. These lasers include carbon dioxide and erbium:yttrium-aluminum-garnet (Er:YAG). This was first demonstrated by Fine et al.[12] in 1967 using a continuous wave CO_2 laser, but the high energy used caused charring of the surrounding corneal tissues. In 1986 Wolbarsdt and coworkers[13] demonstrated another type of laser, the Er:YAG, that can produce a photovaporization effect in the cornea with greater precision and less thermal damage (Figure 1.1.B).

Corneal Photodisruption

Photodisruption occurs when very high laser energy is delivered in an extremely short period of time to a very small space, causing a microscopic explosion.[14] As the laser energy is delivered in a short burst to a small spot, optical breakdown occurs: ionization of the target tissue and subsequent plasma formation.[15] Krasnov[14,16] was the first to use the Q-switched ruby laser in ophthalmology, demonstrating optical breakdown on nonpigmented and transparent structures.

Because the Q-switched ruby laser was not the most efficient delivery system, the search for a better photodisrupter proceeded, including investigation of neodynium (Nd) lasers using both neodymium:glass and Nd:YAG media.[15] Fankhauser, Aron-Rosa, and others pioneered the use of the Nd:YAG laser for posterior capsulotomy,[17,18] peripheral iridotomy,[17–19] cutting of pupillary membranes, and synechiolysis.[18–20] A method of increasing laser efficiency is mode locking. In this method, a shutter within the laser tube allows summation of the oscillating axial modes of light energy and subsequent release of these added pulses in a periodic high-peak power pulse. Q-switched and mode-locked lasers have increased efficiency and efficacy, making photodisruption more accessible for use in ophthalmology.[15] Extremely short pulse photodisruptions paved the way for the newer picosecond photodisrupters designed for intrastromal photorefractive keratectomy.[21,22]

When the laser pulse interacts with corneal tissue, it generates a laser-induced plasma that consists of localized high-temperature ionized gas generated by the high energy absorption. The high temperatures associated with plasma formation result in localized heating of the tissue and production of a gas-liquid phase change, giving rise to cavitation or gas bubble formation.[23–25]

The histopathological studies of intrastromal PRK reveal little thermal damage of tissue adjacent to the site of cavitation. Mass spectroscopical analysis of the cavitation bubbles reveals a mixture of CO_2, water vapor, and CO.[26]

Photothermal Action on the Cornea

The use of heat energy to reshape the corneal surface, known as *thermokeratoplasty*, was studied in the mid-1970s using thermal probes to produce corneal tissue shrinkage.[27–30] Peyman and coworkers later used a CO_2 laser to modify corneal curvature, noting it to be ineffective due to the shallow penetration depth of the laser beam.[31]

Horn et al.[32] in 1990 used a cobalt-magnesium-fluoride laser in rabbit corneas and demonstrated that its effect was fairly stable.

In 1990 Seiler et al.[33] used a holmium laser with a contact probe application (400-μm fiberoptic). They showed that hyperopic correction is achievable in human corneas using this type of thermal energy and is inversely related to the distance of the coagulations from the center of the cornea.

Corneal Photoablation

Srinivasan and coworkers[34] proposed that the photoablation in polymer materials, such as polymethyl methacrylate (PMMA), by ultraviolet (UV) laser pulses can be mainly attributed to a photochemical mechanism. They emphasized that a threshold for ablative decomposition of plastics may exist. At 160 mJ/cm^2 per pulse, an etch mark was clearly visible in reflection, whereas 40 mJ/cm^2 did not leave a mark.[34] This suggested the possibility of optical fine tuning of the ablated surface, and the application of this modality to biologic materials. This was demonstrated by uses such as ablation of fine groves in a strand of human hair.[35,36]

Taboda and coworkers[37,38] were the first to report the ablative effect of KrF, and later of ArF, on rabbit corneas in 1981. Following that report, Trokel and colleagues achieved a precise linear ablation of bovine cornea with 193-nm radiation. Histologic analysis showed no collateral damage and a uniquely smooth ablated surface.[39]

The idea of direct surface recontouring emerged in 1984 when Munnerlyn and coworkers removed central tissue from the anterior surface of the cornea. Only 5 μm of tissue had to be removed to lower the corneal refractive power of a 4-mm optical zone by 1 D. They introduced the term *photorefractive keratectomy* (PRK) in a paper, which was later published in 1988.[40]

Investigators have since been working on both therapeutic and refractive uses of the excimer laser. To proceed with the application of excimer lasers, one must understand the unique action of these UV lasers and the factors helping or hindering their optimum operation.

OPTICS AND MACHINERY DESIGN OF EXCIMER LASER

The excimer laser produces invisible light in the UV sector of the electromagnetic spectrum. The wavelength and photon energy are determined by the gas mixture in the laser cavity. Only ArF, KrF, XeCL, and XeF have sufficient energy in their outputs for practical surgical applications.[40] Krueger and colleagues[41] demonstrated the spectrum of UV tissue ablation by comparing the tissue effects of the four major wavelengths produced by excimer lasers. By varying the type of excited gas in the laser cavity, a different wavelength is generated with a different photon energy. An ArF gas mixture is used to generate the 193-nm wavelength. A KrF gas mixture produces a wavelength of 248 nm, a XeCl gas mixture produces a wavelength of 308 nm, and an XeF gas mixture produces a wavelength of 351 nm. Of these, only the 193-nm wavelength excimer laser produced by the ArF gas mixture and the 308-nm wavelength excimer laser produced by the XeF gas mixture have found medical applications.[40] The energy of the excimer laser photon exceeds the 3.5 eV necessary for breakage of C-C bonds.

All current excimer lasers use an operating microscope to align the laser beam and deliver energy to that portion of the cornea which surrounds the entrance of the pupil of the eye. The rectangular output of the laser cavity is folded, passed through homogenizing optics, and rotated to produce radial symmetry. An iris diaphragm and an adjustable slit are computer-controlled to adjust the distribution of light on the corneal surface.

Laser Principles

Like all lasers, the excimer has several basic components: an energy source, an active medium, and an optical cavity. Before lasing activity can begin, a "population inversion" (more molecules in an excited state than a ground state) must be achieved by pumping the active medium (ArF gas) with a high-energy electrical discharge. The discharge is made with a pair of electrodes across the active medium and perpendicular to the direction of the laser beam (Figure 1.2). In practice, the active medium of ArF is buffered with helium or neon gas, which allows for efficient transfer of the electrical energy to the other two gases.[41]

Argon and fluoride are usually not associated in a single molecule at lower energy levels but do have an affinity for each other at higher levels (Figure 1.3). Therefore, if the ArF gas is pumped properly, it is not difficult to attain a very large population inversion, giving the laser a very high gain.

At this point, an excited (and unstable) ArF molecule can decay spontaneously to a lower state, emitting a photon with a wavelength of 193 nm and initiating a chain reaction by other excited molecules. The mirrors which define the optical cavity on either end of the active medium produce the feedback necessary to sustain the chain reaction, and

Excimer Laser Optics and Corneal Applications

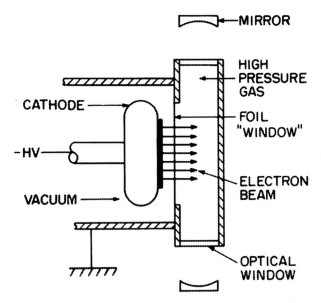

Figure 1.2. Schematic illustration of the ArF laser showing the electron beam discharge across the electrodes (arrows) perpendicular to the direction of the laser beam. (From Rhodes CK, *Topics in Applied Physics: Excimer Laser.* Berlin, Springer-Verlag, 1984.)

lasing action is achieved. One mirror is totally reflective and the other is partially reflective, allowing for emission of laser light. The high gain of the excimer allows for a relatively low reflectance of 4% for the mirror which transmits the laser radiation. The short lifetimes of the excited ArF molecules translate into very short pulses of laser light (on the order of nanoseconds). This is useful in delivering UV energy to corneal tissue quickly, minimizing thermal effects.

Compared to many other lasers, an excimer laser beam profile is not very uniform in intensity. Since even rather small variations in intensity can result in unacceptably large differences in ablation across the cornea, measures must be taken to smooth the beam. In some designs, this is accomplished first by passing the beam through an aperture to eliminate variations along the edges. Next, the rectangular beam is expanded to a square shape using a pair of prisms. A "K-mirror"-type beam rotator is then used to time-average the pulsed energy and ensure a smooth, radially symmetric profile of beam intensity (Figure 1.3).

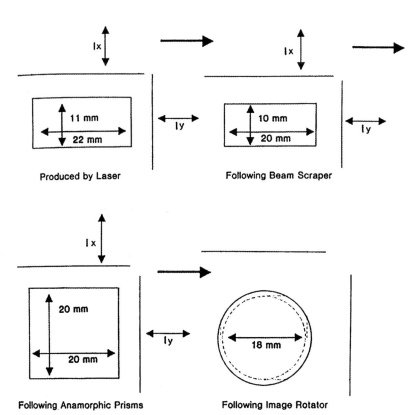

Figure 1.3. Excimer laser beam intensity cross section as it emerges from the laser. (From L'Esperance et al, *Arch Ophthalmol* 1989;107:133.)

Several ideas are useful for optimizing excimer laser performance. Accurate alignment of both the laser cavity and beam delivery system optics is aided by the use of a helium-neon laser. Mirrors are coated for best reflection at the working wavelength, and selected optics are made of CaF or MgF to increase transmission. (The optics of the laser are easily degraded by the high energy output and must be checked regularly.) The gases in the active medium will decay over a relatively short time period, requiring close monitoring and replenishment to achieve stable, uniform pulses.

Patient Considerations

The laser beam is now in a form which is suitable for use in corneal surgery. Before treatment of refractive errors is considered, however, modification of the beam is necessary. For myopic corrections, this can be achieved with either an expanding iris diaphragm (allowing for more tissue removal centrally) or a series of apertures on a "myopic wheel" which are stepped into place as the procedure proceeds. Comparable wheels are available for hyperopic and astigmatic corrections (Figures 1.4, 1.5, and 1.6).

Before the beam exits the laser, it passes through a safety shutter and a lens which focuses the beam in front of the cornea. The latter feature ensures that diverging laser light falls upon the cornea, preventing the tissue from being exposed to too high an intensity. Some of the beam is split off before it reaches the patient to provide measurements and monitoring for the laser.

An alignment subsystem is needed to ensure accurate centration of the cornea during treatment. The ophthalmologist uses a binocular microscope with a reticle pattern to align the laser and focus it properly on the cornea. The patient is asked to fixate on a light in the system during the procedure, and his or her compliance is monitored.

EXCIMER LASER MECHANISM OF ACTION: ABLATIVE DECOMPOSITION

When the energy density of laser light exceeds a critical value, molecular bonds are broken in a process known as *photoablation*. The energy of a photon of light at 193 nm is 6.4 eV. This is greater than that of carbon–carbon (3.5 eV) or peptide (3.0 eV) bonds. Proteins in the cornea have a high absorption band near 190 nm,[41-43] making the use of an ArF laser especially advantageous. Because of the high absorption of the UV laser light by corneal tissue, the photon energy is concentrated in a very small volume. The intense buildup in this volume in an ex-tremely short time (picoseconds) results in the breakdown of polymer chains into small fragments which are ejected at high speed (>1000 m/s) from the surface of the cornea.[44] Since this photoablative process occurs on a very short time scale,

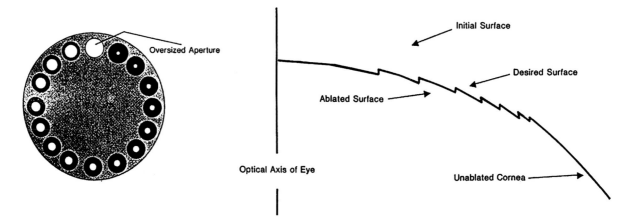

Figure 1.4. Left: Schematic illustration of an aperture wheel that allows the laser beam to ablate the central cornea more deeply than the peripheral cornea to increase the radius of curvature of the anterior corneal surface to reduce myopia. Right: Representation of the stepped profile ablation of the cornea to reduce myopia. (From L'Esperance et al, *Arch Ophthalmol* 1989;107:133.)

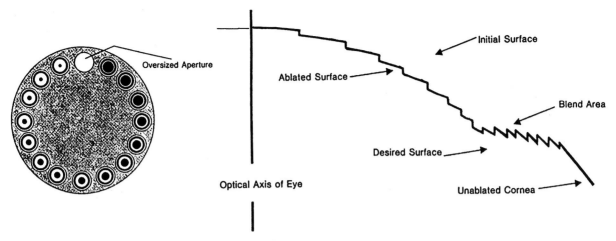

Figure 1.5. Left: Schematic illustration of an aperture wheel that allows the laser beam to ablate the peripheral cornea more deeply than the central cornea to decrease the radius of curvature of the anterior corneal surface in order to reduce hyperopia. Right: Representation of the stepped profile ablation of the cornea to reduce hyperopia. (From L'Esperance et al, Arch Ophthalmol 1989;107:134.)

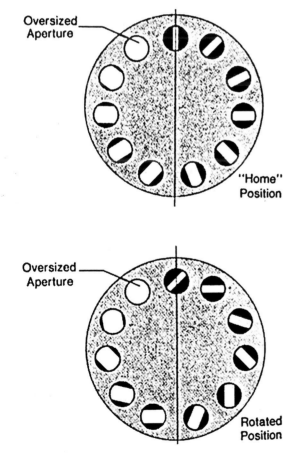

Figure 1.6. Schematic illustration of the astigmatism aperture wheel. (From L'Esperance et al, Arch Ophthalmol 1989;107:134.)

there is little transfer of energy to adjacent tissue and thermal effects are minimal[45] (especially at repetition rates of several Hertz). However, it is possible that some of the tissue fragments may settle back onto the cornea, causing a slight decrease in the ablation rate of subsequent pulses.

Ablation Threshold, Rates, and Depth

Each pulse of laser energy removes a precise amount of corneal tissue of uniform depth.[41] The quantitative studies of excimer action demonstrated a threshold phenomenon for ablation at all excimer wavelengths. At low irradiance levels, the cornea remains grossly unaffected by laser energy. As irradiance of the laser exposure increases, a faint clouding of the exposed corneal tissue occurs, ranging from a faint discoloration at 193 nm and 248 nm to a penetrating stromal clouding at 308 nm and 351 nm. With a further increase, coagulative changes are seen at all excimer wavelengths except 193 nm, suggesting that at 193 nm an ablation is purely photochemical (and the thermal component is absent), while a thermal component exists for other UV wavelengths.[46] The threshold for ablation at 193 nm is 50 mJ/cm^2 and is lowest among the available excimer laser wavelengths. With longer wavelengths the ablation threshold increases logarithmically.[46] The corneal ablation depth depends on the amount of energy striking the cornea and follows a roughly sigmoidal curve with increasing pulse irradiance.[41] This curve flattens out at approximately 60 mJ/cm^2 (for 193 nm), so no higher ablation rate is achieved with higher fluences (energy/unit area). The most efficient fluence (in terms of tissue removed per pulse) seems to be about 200 mJ/cm^2 at 193 nm and 1000 mJ/cm^2 at 248 nm.[47] At typical energy densities of 160–200 mJ/cm^2 (at 193 nm), 0.2–0.4 μm of tissue is ablated per pulse. The exact rate is dependent on factors such as corneal thickness and hydration.[48] In practice, excimer laser calibration must be carefully monitored to ensure accurate ablation rates.

Puliafito and coworkers[49] performed comparative experimental studies of excimer laser ablation of the cornea and crystalline lens using both 193-nm (ArF) and 248-nm (KrF) radiation. They reported that the threshold fluence for corneal ablation was higher at 248 nm. Furthermore, they found that incisions made with 248-nm radiation at fluences comparable to those of the 193-nm experiments showed much greater damage (an order of magnitude >2500 nm) to tissue adjacent to the ablated area.[49] These alterations included denaturation of stromal collagen with ultrastructural evidence of marked disorganization. Edges of incisions made with 248 nm were irregular in contrast to the sharp edges seen at 193 nm.[49] Marshall and colleagues,[50] using ultrastructural analysis of the walls of the ablated areas, found that damage to the adjacent structures was confined to a zone of 60 to 200 nm in width when a 193-nm exciser laser was used.

SAFETY

As new types of lasers become practical for medical use, we must study not only the unexpected effects of laser light, but also any other hazards associated with the use of laser energy and devices. This includes the safety of the patient, surgeon, surgical assistants, and environment. For example, fluorine gas is extremely toxic. Important safety factors that must be considered are possible mutagenesis, carcinogenesis, corneal integrity, cataractogenesis, and retinal toxicity by UV light.

Cataractogenesis

Laser-induced fluorescence occurs as the excimer laser light strikes the cornea, emitting a broad spectral band of UV light of longer wavelength. These secondary UV wavelengths may be more deleterious.[51] Wavelengths between 295 and 400 nm are partially transmitted by the cornea and absorbed by the lens. During photoablation, the cornea absorbs 100% of the energy of wavelengths shorter than 280 nm. It absorbs more than 90% of photon energy when the wavelength increases to 300 nm, but about 2% reaches the lens. At 360 nm, the lens absorbs most photons.[52] Tufe and coworkers calculated the percentage of excimer laser radiant energy converted into fluorescent light of longer UV wavelengths and found it to be 0.001%. (Considering 400 pulses at 200 mJ/cm^2 and a 4-mm pupil, the amount of energy delivered to the lens is 10 μJ/cm^2. This value is 2000 to 10,000 times lower than the threshold for generating lens opacities in rabbits.[53] Wavelengths longer than 320 nm are not energetic enough to break bonds, and wavelengths shorter than 300 nm are absorbed by the cornea. So an acute cataract can result from only a narrow band of electromagnetic wavelengths.[52]

Mutagenesis and Carcinogenesis

UV light is known to be both mutagenic and carcinogenic.[54,55] In vitro tests have shown that the use of 193-nm wavelength excimer laser is not associated with significant mutagenesis or carcinogenesis. Both the penetration depth of the radiation and the absorption spectrum of DNA in each cell determine how mutagenic a given wavelength of laser light will be.[56] Trentacoste and colleagues[57] showed that in vitro irradiation with the 193-nm exciser laser does not result in cytotoxic damage to DNA and mutagenicity, as Nuss and colleagues[58] showed using in vivo irradiation. Excision repair appears to be the most important mechanism for removing the damaged DNA, and this can be monitored by the amount of unscheduled DNA synthesis. When studied at the edge of ablated rabbit cornea, no replacement of damaged DNA was shown when the tissue was irradiated with a 193-nm excimer, but significant replacement occurred when the tissue was irradiated with a 248-nm excimer laser.[58]

Heat Generation

Excimer laser photoablation is believed to be predominantly a photochemical process, yet the release of energy as molecular bonds break gives rise to temperature changes on the surface. Several authors[56,59] have reported temperature increases of 10°–20°C in stromal tissue adjacent to the laser-ablated area. High-energy UV radiation breaks protein molecules into diatomic and triatomic fragments which are blown away from the treated area. Because of the tremendous energy, the ablated fragments take with them much of the surplus energy within the system and may thus prevent thermal damage to the surrounding tissue.[60]

Corneal Integrity

Campos and coworkers[61] studied ocular integrity after refractive surgery in porcine eyes. Eyes with radial keratotomy (RK), and phototherapeutic keratectomy (PTK) at PRK (−10 D), various depths were examined. They found that upon globe compression, the eyes that underwent RK ruptured at the incision site. Eyes that underwent PTK had corneal rupture when stromal ablations exceeded 40% of corneal thickness. Eyes that underwent −10 PRK ruptured along the sclera, as did normal eyes.

CORNEAL WOUND HEALING

After excision of the superficial stroma with a 193-nm ArF excimer laser, reepithelialization is usually complete by 3–7 days.[62,63] Over the following 6 months the epithelial thickness increases, especially at the deepest part of the ablated area[64,65] (Figure 1.7). Most studies show that there is minimal risk of epithelial erosion after excimer laser photoablation.[66,67] Recurrent erosions have been observed rarely in patients undergoing PRK.[68,69]

Following corneal excimer ablation, remnants of Bowman's layer have been noted within the ablation site.[66,70] The role of Bowman's layer in maintaining corneal transparency is negligible. This conclusion is based on the presence of normal corneal clarity in animals whose corneas lack a Bowman's membrane, rare reports of cases of congenital absence of Bowman's layer (with maintaining corneal clarity), the lack of recurrent erosions following laser treatment, and excellent morphologic recovery of normal epithelial adhesion complexes.[71]

Fantes and colleagues[62] believe that changes in the anterior stroma are the major cause of postoperative corneal haze. As the corneal stroma responds to the laser injury, new collagen and proteoglycans are produced which replace the early "pseudomembrane" in the subepithelial region. Disorganized collagen is also seen.[70,72]

Several studies have shown no or minimal effect of excimer laser corneal ablation on Descemet's membrane.[62,66,67] Endothelial injury following 193-nm excimer laser corneal ablation was seen only when the ablation was close to Descemet's membrane.[73] Significant endothelial cell loss was reported 1 year following surgery by Beldaus and colleagues.[74]

MAINTENANCE OF OPTICAL CLARITY

The appearance of superficial stromal opacification after corneal ablation with the 193-nm excimer laser is extremely common and represents a biological healing process which is detected clinically during the first postoperative month, peaks at 1–3 months postoperatively, and gradually decreases over the first year[75] (Figures 1.7 to 1.10).

This superficial stromal opacification was described by L'Esperance and coworkers as a reticular haze.[76] Several authors have studied the healing response in animals,[77–82] and in humans, using slit lamp biomicroscopy,[82–85] the opacity lensometer,[86] and confocal microscopy.[86,88] The subjective evaluation of haze using a slit lamp is the most practical way of grading haze following excimer laser treatment. Several authors[62,89–91] have subjectively classified haze as follows: grade 0 (a totally clear cornea), grade 0.5 (barely perceptible haze, seen only by indirect broad tangential illumination), grade 1 (trace haze of minimal density seen with direct and diffuse illumination), grade 2 (mild haze, easily visible with direct focal illumination), grade 3 (moderate haze that partially obscures iris details), and grade 4 (severe haze that completely obscures iris details).

True haze is that experienced by the patient as scattered light, yet visually significant haze can be roughly estimated when slit lamp assessment shows haze grading of 1.5 on the classification scale. Visually significant haze can be seen in approximately 3–11% of patients[82,90] and in 10–40% of those with the correction of high degrees of myopia.[90,92]

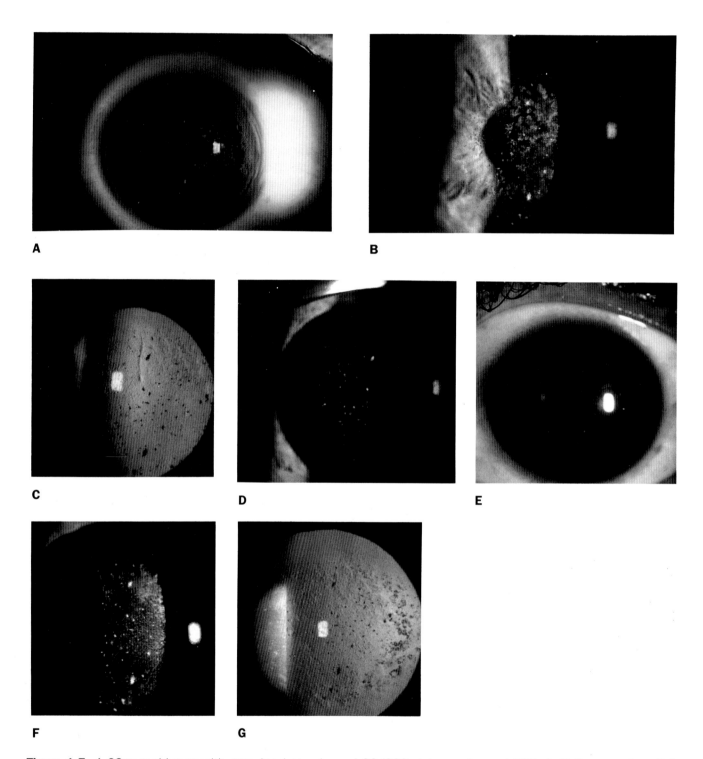

Figure 1.7. A 38-year-old man with granular dystrophy and 20/200 vision underwent PTK. *A, B:* Preoperative clinical appearance by retroillumination and oblique illumination, respectively, demonstrating the lesion in the left eye. *C:* Retroillumination of the left eye 3 months after PTK. *D:* Slit lamp illumination of the left eye 3 months after PTK. Visual acuity improved to 20/30 3 months after PTK. *E–G:* Right eye of the same patient 6 months after PTK under a different lighting condition. His vision improved to 20/30.

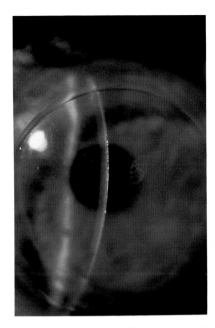

Figure 1.8. Preoperative slit lamp photograph with oblique illumination of a 45-year-old man with lattice dystrophy. The patient suffered a marked decrease in vision, with 20/1000 visual acuity, and improved to 20/100 in 24 months.

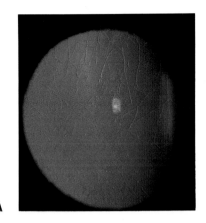

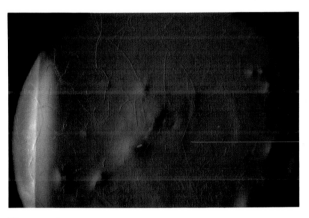

Figure 1.10. Retroillumination of both eyes of a 35-year-old man with lattice dystrophy showing *(A):* the right eye 14 months after PTK and *(B):* the unoperated left eye.

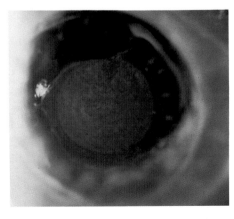

Figure 1.9. A 1-day postoperative slit lamp photograph. The patient was treated for high myopia. Scleral scatter shows the 5.5-mm treatment area.

Studies have shown that high levels of correction,[82,92] smaller ablation diameters (less than 4.5 mm),[90] male gender,[90] and poor compliance with the postoperative steroid medication regimen[82,90] are some of the factors that correlate with increased density of haze.

Steroids have proven effective in reducing postoperative haze.[77,82,90] Investigators studied the effects of some other drugs in reducing corneal haze, such as mitomycin,[93] protease inhibitors such as aprotnin,[83] and interferon-alpha 2b.[94]

EXCIMER LASER APPLICATIONS FOR REFRACTIVE SURGERY

Correction of Myopia

Laser RK

Laser RK follows the approach that has been adopted for RK for the correction of myopia and astigmatism. Because the excimer laser produces an excision rather than an incision (unlike that of a diamond knife), the ingrowth of a broader epithelial plug might result in slower wound healing and thereby not only an increased or unpredictable result (compared to that achieved with conventional RK), but also prolonged diurnal fluctuations and diminished long-term stability of the achieved correction.[95]

Photorefractive Keratectomy

The idea of direct surface recontouring proposed by Munnerlyn and coworkers[40] consisted of removing central corneal tissue from the anterior surface of the cornea. Eventually this became the basis of PRK for the correction of myopia.

In 1987, the first superficial keratectomies on human subjects performed by excimer laser photablation on blind eyes or eyes destined for enucleation, were performed by L'Esperance and colleagues.[96] The first successful excimer laser PRK for myopia on a normally sighted human eye was performed in June 1988 by McDonald and colleagues. This encouraged the Food and Drug Administration (FDA) to allow the VISX and Summit excimer lasers to undergo a controlled study. The results for both have been completed under the stringent guidelines issued by the FDA for patient selection and clinical evaluation of this new technique in the United States.

Correction of Astigmatism

In initial attempts to correct corneal astigmatism using a 193-nm excimer laser, where the laser produced curved excisions that essentially mimic the classic T-incisions made with a knife, Seiler and coworkers[97] reported the correction of up to 4.16 D of corneal astigmatism using the suction cup slit technique. In 1989 Lang and coworkers reported on their experiments to treat corneal astigmatism through a elliptic ablabion zone.[98] Later, McDonnell and coworkers[99,100] studied this toric approach to corneal ablation on PMMA blocks, plastic coreas, rabbit corneas, and human corneas. They decided that these elliptical ablation techniques represent a new approach to the correction of corneal astigmatism.

Correction of Hyperopia

Surgical correction of hyperopia is more difficult than surgical correction of myopia. However, many techniques have been developed to correct hypermetropia, including lamellar techniques such as keratomileusis, keratophakia and epikeratophakia, thermokeratoplasty, transverse keratotomy, hexagonal keratotomy, and PRK.[101] Del Pero and coworkers[102] reported their results on six cynomolgus monkeys treated with a 193-nm excimer laser. They achieved a 5.2-D correction at 1 year of follow-up, with an intended correction of 6 D.

PRK for correction of hyperopia appears to be safe and effective in the short term. Evaluation of more eyes with longer follow-up should provide a clearer idea of the safety and efficiency of this mode of treatment.

Preparation of Plano or Powered Lenticules for Epikeratophakia

Epikeratophakia, originally described by Barraquer, requires freezing of the corneal lenticule to reshape it. This causes the death of all keratocytes during the procedure unless cryoprotective measures are taken.[103]

Gabay and colleagues[104] demonstrated the laser-processed lenticule and successfully transplanted it into a human eye.

Phototherapeutic Keratectomy

PTK is the removal of anterior layers of a diseased cornea using the excimer laser. Excimer laser photoablation can etch corneal tissue precisely, leading to a smooth, uniform surface covered by a pseudomembrane.[39,49,50] This smoothness and uniformity seem to contribute to the maintenance of corneal transparency and the absence of significant scar formation. It has long been believed that Bowman's layer was essential in maintaining corneal transparency. This belief may be due to the use of conventional surgical techniques to change corneal shape, which could not produce the smooth, uniform surface produced by the laser.[105,106] PTK underwent FDA-sponsored multicenter trials to assess its safety and therapeutic effectiveness in human corneas.

Several reports have shown successful results of PTK in posttraumatic corneal erosions. Forger et al.[107] reported the successful management of posttraumatic corneal erosions in all of their nine patients. Sher et al.[108] reported the successful treatment of only one of two patients treated for posttraumatic cornea erosion.

McDonnell and Seiler[109] reported successful management of Reis-Bücklers' corneal dystrophy using the 193-nm excimer laser in two patients.

In the management of corneal scars, Campos et al.[110] reported significant visual improvement in two of six patients with postinfectious scars; by contrast, none of their three patients with posttraumatic scars improved. Stark and coworkers[111] reported significant improvement in visual function in four of seven patients with scars limited to the superficial stroma, while Sher et al.[108] reported significant visual improvement in three of five patients with postinfectious corneal scars and visual improvement in five of eight patients with posttraumatic scars. Several investigators have reported on the treatment of posttraumatic nodules and scars, postinfectious scars, corneal dystrophies, corneal surface irregularities, and ablation of infectious crystalline keratopathy.[107–114]

In addition to the side effects previously mentioned, hyperopia appears to be the most annoying side effect with excimer laser PTK. Chamon et al.[114] observed a positive correlation between depth of stromal ablation and amount of induced hyperopia. Campos et al.[110] reported a hyperopic shift in 55% of 18 treated eyes.

The 193-nm ArF excimer laser was also used to perform prismatic photoablation in model eyes and in rabbit corneas.[115] Experiments were conducted to determine the feasibility of performing prismatic photoablation and its potential in correcting strabismus and diplopia. Prismatic photoablation of PMMA blocks and lenses were not associated with refractive changes accompanying the prismatic effect. Prismatic photokeratectomy (PPK) has great potential for fine-tuning the results of conventional strabismus surgery and for patients with stable diplopia following nerve palsy and ocular surgery.

The ArF excimer laser system is ideal for both refractive and therapeutic corneal surgery. Its reliable performance and beam uniformity ensure that minimal surface irregularities are produced in the ablation zone. The excimer radiation is very highly absorbed in the first several microns of the cornea, ensuring safety and precision. The very short pulse length produces minimal thermal damage to adjacent tissues. Options are available for the treatment of myopia, hyperopia, astigmatism, and superficial corneal disease.

REFERENCES

1. Cherry PM, Faulkner JD, Shaver RP, et al: Argon laser treatment of corneal neovascularization. *Ann Ophthalmol* 1973;5:911–920.
2. Velasco JE, Sester DW: Bound free emission spectra of diatomic xenon halides. *J Chem Phys* 1975;62:1990–1991.
3. Seades SK, Hart GA: *Appl Phys Lett* 1975;27:263.
4. Ewing JJ, Brau CA: *Appl Phys Lett* 1975;27:350.
5. Brau CA, Ewing JJ: *Appl Phys Lett* 1975;27:435.
6. Hoffman JM, Hays AK, Tisone GC: High-power UV noble gas halide laser. *Appl Phys Lett* 1976;28:538–539.
7. Stevens B, Hutton E: Radiative lifetime of the pyrene dimer and the possible role of excited dimers in energy transfer processes. *Nature* 1960;186:1045–1046.
8. Bumham R, Djeu N: Ultraviolet preionized discharge-pumped lasers in Xef, Krf, Arf. *Appl Phys Lett* 1976;29:707–709.
9. Marsh RJ, Marshall J: Treatment of lipid keratopathy with the argon laser. *Br J Ophthalmol* 1982;66:127–135.
10. Bahn CF, Vine AK, Wofter JR, et al: Argon laser photocoagulation of a vitreocorneal adhesions after trauma. *Ophthalmic Surg* 1982;13:53–55.
11. Baer JC, Foster CS: Corneal laser photocoagulation for the treatment of neovascularization. Efficacy of 577 nm yellow dye laser. *Ophthalmology* 1992;99:173–179.
12. Fine BS, Fine S, Peacock GR, et al: Preliminary observations on ocular effects of high power, continuous CO_2 laser irradiation. *Am J Ophthalmol* 1967;64:209–222.
13. Wolbarsdt ML, Fowlks GN, Esterwitz L, et al: Corneal surgery with an ER:YAG laser at 2.94 nm. *Invest Ophthalmol Vis Sci* 1986;27(Suppl):93.

14. Krasnov MM: Laseropuncture of the anterior chamber angle in glaucoma. *Am J Ophthalmol* 1973; 25:674–678.
15. Steinart RF, Puliafito CA: *The ND:YAG Laser in Ophthamology: Principles and Clinical Application of Photodisruptors.* Philadelphia, WB Saunders, 1985, pp 5–21.
16. Krasnov MM: Q-switched laser goniopuncture. *Arch Ophthalmol* 1979;92:37–41.
17. Aron-Rosa D, Aron J, Griesemann M, et al: Use of ND:YAG laser to open the posterior capsule after lens implant surgery: A preliminary report. *Am Intraocul Implant Soc J* 1980;6:532–534.
18. Fankhauser F, Roussal P, Steffen J, et al: Clinical studies on the efficiency of high power laser radiation upon some structures of the anterior segment of the eye. (First experience of the treatment of some pathological conditions of the anterior segment of the human eye by means of Q-switched laser system). *Int Ophthalmol Clin* 1981;3:129–139.
19. Fankhauser F: The Q-switched laser, principles and clinical results. In Trokel SL (ed): *YAG Laser Ophthalmic Microsurgery.* Norwalk, CT, Appleton-Century-Crofts, pp 101–146.
20. Klapper RM: Q-switched ND:YAG laser iridotomy. *Ophthalmology* 1984;91:1017–1021.
21. Fankhauser F, Rol P: Microsurgery with ND:YAG laser, an overview. *Int Ophthalmol Clin* 1985;25: 55–58.
22. Niemz MH, Kanchik EG, Bille JF: Plasma-mediated ablation of corneal tissue at 1053 nm using a ND:YAG oscillator/regenerative amplifier laser. *Lasers Surg Med* 1991;11:426–431.
23. Niemz MH, Hoppeler TP, Juhasz T, et al: Intrastromal ablations for retractive corneal surgery using picosecond infrared laser pulses. *Lasers light Ophthalmol* 1993;5:149–155.
24. Lin CP: Laser tissue interactions: Basic principles. *Ophthalmol Clin North Am* 1993;6:381–391.
25. Fujimoto JG, Lin WZ, Ippen EP, et al: Time resolved studies of ND:YAG laser-induced breakdown plasma formation, acoustic wave generation and cavitation. *Invest Ophthalmol Vis Sci* 1985;26: 1771–1777, 1985.
26. Zysset B, Fujimoto JG, Puliafito CA, et al: Picosecond optical breakdown: Tissue effects and reduction of collateral damage. *Lasers Surg Med* 1989; 9:193.
27. Kaiser R, Habib MS, Speaker MG, et al: Mass spectrometry analysis of cavitation bubbles generated by intrastromal ablation with the ND:YLF picosecond laser. *Invest Ophthalmol Vis Sci* 1994;35(Suppl): 2020.
28. Gasset A, Shaw E, Kaufman H, et al: Thermokeratoplasty. *Trans Am Acad Ophthalmol Otolaryngol* 1973;77:441–451.
29. Aquavella J: Thermokeratoplasty. *Ophthalmic Surg* 1974;4:39–48.
30. Keates R, Dingle J: Thermokeratoplasy for keratoconus. *Ophthalmic Surg* 1975;6:89–92.
31. Peyman GA, Larson B, Raichard M, et al: Modification of rabbit corneal curvature with use of carbon dioxide laser burns. *Ophthalmic Surg* 1980; 11:325–329.
32. Hom G, Spears KG, Lopes O, et al: New refractive method for laser thermal keratoplasty with the CO:MsF$_2$ laser. *J Cataract Refract Surg* 1990;16: 611–616.
33. Seiler T, Matallana M, Bende T: Laser thermokeratoplasty by means of a pubed holmium:YAG laser for hyperopic correction. *Refract Corneal Surg* 1990; 6:355–359.
34. Srinivasan R, Baren B, Dreyfus RW, et al: Mechanism of the ultraviolet laser ablation of polymethyl methacrylate at 193 nm and 248 nm: Laser induced fluorescence analysis, chemical analysis and doping studies. *J Opt. Soc Am* 1986;3:785–791.
35. Sutcliffe E, Srinivasan R: Dynamics of UV laser ablation of polymers surfaces. *J Appl Phys* 1986; 60:3315–3322.
36. Srinivasan R: Kinetics of ablative photodecomposition of organic polymers in the far-ultraviolet (193 nm). *J Vac Sci Technology B* 1983;1:923–926.
37. Taboada J, Mikesell GW, Reed RD: Response of the corneal epithelium to KrF excimer laser pulse. *Health Phys* 1981;40:677–683.
38. Taboada J, Archibald CJ: An extreme sensitivity in the corneal epithelium to far UV Arf excimer laser pulses. Procedures of the Scientific Program of the Aerospace Medical Association, San Antonio, 1981.
39. Trokel SL, Srinivasan R, Baren B: Excimer laser surgery of the cornea. *Am J Ophthalmol* 1983;46: 710–715,
40. Munnerlyn CR, Koors SJ, Marshall J: Photoretractive keratectomy: A technique for laser refractive surgery. *J Cataract Refract Surg* 1988;14:46–52.
41. Waring GO: *Refractive Keratotomy for Myopia and Astigmatism.* St Louis, Mosby-Year Book, 1992, p 689.
42. Azar DT, Hahn TW, Khoury JM: Corneal wound healing following laser surgery. In *Refractive Surgery,* Azar DT (ed), Appleton and Lange, 1977, Stamford, CT, pp 41–74.
43. Ren Q, Simon G, Pavel JM: Ultraviolet solid state

laser photorefractive keratectomy. *Ophthalmology* 1993;100:1828–1834.
44. Waring GO: Development of a system for excimer laser corneal surgery. *Trans Am Ophthalmol Soc* 1989; 87:875.
45. Krauss JM, Pulifito CA: Lasers in ophthalmology. *Lasers Surg Med* 1995;17:128.
46. Krueger RR, Trokel SL, Schubert HD: Interaction of UV laser light with the cornea. *Invest Ophthalmol Vis Sci* 1985;26:1455–1464.
47. Krueger RR, Trokel SL: Quantitation of corneal ablation by ultraviolet laser light. *Arch Ophthalmol* 1985;103:1741.
48. Waring GO: Development of a system for excimer laser corneal surgery. *Trans Am Ophthalmol Soc* 1989;87:881.
49. Puliafito CA, Steinert RF, Deutsch TF, et al: Excimer laser ablation of the cornea and lens. *Ophthalmology* 1985;92:741–748.
50. Marshall J, Trokel S, Rothery S, et al: An ultrastructural study of corneal incisions induced by an excimer laser at 193 nm. *Ophthalmology* 1985;92:749–758.
51. Bends T, Seiler T, Woliensak J: Side effects in excimer laser corneal surgery. Corneal thermal gradients. *Graefes Arch Clin Exp Ophthalmol* 1988; 226:277–280.
52. Bores LD: *Refractive Eye Surgery*. Boston, Blackwell, 1993, p 405.
53. Tuft S, Al-Dhahir R, Dyer P, et al: Characterization of the fluorescence spectra produced by the excimer laser radiation of the cornea. *Invest Ophthalmd Vis Sci* 1990;108:915–916.
54. Burton JL: Accuracy of local data on skin cancer. *Br Med J* 1995;310;1328.
55. Liu M, Pelling JC: UV-B/A irradiation of mouse keratinocytes results in p53-mediated WAF1/CIP1 expression. *Oncogene* 1995;10:1955–1960.
56. Seiler T, Bende T, Winckler K, et al: Side effects in excimer laser corneal surgery: DNA damage as a result of 193 nm excimer laser radiation. *Graefes Arch Clin Exp Ophthalmol* 1988;226:273–276.
57. Trentacoste J, Thompson K, Parrish RK, et al: Mutagenic potential of 193 nm excimer laser on fibroblasts in issue culture. *Ophthalmology* 1987; 94:125–129.
58. Nuss RC, Puliafito CA, Dehn E: Unscheduled DNA synthesis following excimer laser ablation of the cornea in vivo. *Invest Ophthalmol Vis Sci* 1987;28: 287–294.
59. Beins MW, Uaw LH, Oliva A, et al: An acute light and electron microscopic study of UV 193 nm excimer laser corneal incisions. *Ophthalmology* 1988; 95:1422–1433.
60. Puliafito CA, Stem D, Krueger RR, et al: High speed photography of excimer laser ablation of the cornea. *Arch Ophthalmol* 1987;105:1255–1259.
61. Campos M, Lee McDonnell PJ: Ocular integrity after refractive surgery: Effects of photorefractive keratotomy, phototherapeutic keratotomy and radial keratotomy. *Ophthalmic Surg* 1992;23: 598–602.
62. Fantes FE, Hanna KD, Waring GO III, et al: Wound healing after excimer laser keratomileusis (photorefractive keratectomy) in monkeys. *Arch Ophthalmol* 1990;108:665–675.
63. Seiler T, Woliensak J: Myopic photorefractive keratectomy with excimer laser. One year follow-up. *Ophthalmology* 1993;98:1156–1163.
64. Campos M, Cuevas K, Shieh E, et al: Corneal wound healing after excimer laser ablation in rabbits, expanding versus contracting apertures. *Refract Corneal Surg* 1992;8:378–381.
65. Renard G, Hanna K, Saragoussi JJ, et al: Excimer laser experimental keratectomy: Ultrastructural study. *Cornea* 1987;6:269–272.
66. Wilson CS, Stark WJ, Green RW: Corneal wound healing after 193 nm excimer laser keratectomy. *Arch Ophthalmol* 1991;109:1426–1432.
67. Verbec MP, McDonald MB, Chase DS, et al: Traumatic corneal abrasions after excimer laser keratectomy. *Am J Ophthalmol* 1993;116:101–102.
68. Basin M, Meller D: Corneal epithelial dots following excimer laser PRK. *J Refract Corn Surg* 1994; 10:357–359.
69. Shieh E, Moveira M, D'Arcy J, et al: Quantitative analysis of wound healing after cylindrical and spherical excimer laser ablations. *Ophthalmology* 1992;99:1050–1055.
70. Binder PS, Anderson J, Rock M, et al: The morphologic features of human excimer laser photoablation. *Ophthalmology* 1994;101:979–989.
71. Salz JJ, McDonnell PJ, McDonald MB: *Corneal Laser Surgery*. St Louis, CV Mosby, 1995, p 79.
72. Marshall J, Trokel S, Rothery S, et al: A comparative study of corneal incisions induced by diamond and steel knives an two ultraviolet radiations from an excimer laser. *Br J Ophthalmol* 1986;70:482–501.
73. Dehm EJ, Puliafito CA, Adier CM: Endothelial injury in rabbits following excimer laser ablation at 193 and 248 nm. *Arch Ophthalmol* 1986;104: 1364–1368.

74. Beidaus R, Thompson K, Waring GO, et al: Quantitative specular microscopy after PRK. *Ophthalmology* 1992;99:125.
75. Marshall J. Trokel SL, Rothery S, Krueger RR. Long term healing of the central cornea after PRK using an excimer laser. *Ophthalmology* 1988;95: 1411–1421.
76. L'Esperance FA, Taylor DM, Warner JW: Human excimer laser keratectomy: Short term history. *J Refract Surg* 1988;4:118–124.
77. Tuft SJ, Zabel RW, Marshall J: Corneal repair following keratectomy. A comparison between conventional surgery and laser photoablation. *Invest Ophthalmol Vis Sci* 1989;30:1769–1777.
78. Campos M, Cuevas K, Shieh E, et al: Corneal wound healing after excimer laser ablation in rabbits: Expanding versus contracting apertures. *Refract Corneal Surg* 1992;8:378–381.
79. Malley D, Steiner R, Puliafito CA, et al: Immunofluorescence study of corneal wound healing after excimer laser anterior keratectomy in the monkey eye. *Arch Ophthalmol* 1990;108:1316–1322.
80. Hanna K, Pouliquen Y, Waring GO III, et al: Corneal wound healing in monkeys after repeated excimer laser keratectomy. *Arch Ophthalmol* 1992; 110:1286–1291.
81. SundarRaj N, Greiss M III, Fantes F, et al: Healing of excimer laser ablated monkey cornea: An immunohistochemical evaluation. *Arch Ophthalmol* 1990;108:1604–1610.
82. Seiler T, Woliensak J: Myopic photorefractive keratectomy with the excimer laser. One year follow-up. *Ophthalmology* 1991;8:1156–1163.
83. Lehmann CP, Obart D, Patmore A, et al: Plasmin inhibitors and plasminogen-activator inhibitors after excimer laser PRK: New concept in prevention of myopic regression and haze. *Invest Ophthalmol Vis Sci* 1993;34:705.
84. Marshall J, Trokel SL, Rothery S, et al: Long term healing of the central cornea after photorefractive keratectomy using an excimer laser. *Ophthalmology* 1988;95:1411–1421.
85. Lehmann CP, Gurtry DS, Kerr Muri M, et al: Corneal haze after excimer laser refractive surgery: Objective measurements and functional implications. *Eur J Ophthalmol* 1991;1:173–180.
86. Angott-Andrade H, Mcdonald M, Un J, et al: Evaluation of an opacity lensometer for determining corneal clarity following excimer laser photoablation. *Refract Corneal Surg* 1990;6:346–351.
87. Haight D, Auran J, Koester C, et al: In vivo confocal scanning sift-microscopy of the human cornea after excimer laser photorefractive keratectomy. *Invest Ophthalmol Vis Sci* 1993;34:703.
88. Rajpal R, Essepian J, Rapnano C, et al: Evaluation of human corneal wound healing with confocal microscopy after excimer laser keratectomy. *Ophthalmology* 1992;99:176.
89. Gurtry DS, Kerr Muri M, Marshall J: Excimer laser photorefractive keratectomy: 18 month follow-up. *Ophthalmology* 1992;8:1209–1219.
90. Caubert E: Cause of subepithelial cornea haze over 18 months after photorefractive keratectomy for myopia. *Refract Corneal Surg* 1993;9(Suppl): 565–570.
91. Seiler T, Kahl G, Kriegerowski M: Excimer laser (193 nm) myopic keratomileusis in sighted and blind human eyes. *Refract Corneal Surg* 1990; 6:165–173.
92. Heitzman J, Binder P, Kassar B, et al: The correction of high myopia with the excimer laser. *Arch Ophthalmol* 1993;111:1627–1634.
93. Talamo J, Goilamudi S, Green W, et al: Modulation of corneal wound healing after excimer laser keratomileusis using topical mitomycin C and steroids. *Arch Ophthalmol* 1991;109:1141–1146.
94. Model N, Gillis MC, Crouch R, et al: Effects of topical interferon-alpha 2b on corneal haze after excimer laser photorefractive keratectomy in rabbits. *Refract Corneal Surg* 1993;9:443–451.
95. Steinert RF, Puliafito CA: Laser corneal surgery. *Int Ophthalmol Clin* 1988;28:150–154.
96. L'Esperance FA Jr, Taylor DM, Warner JW: Human excimer laser keratectomy: Short term histopathology. *J Refract Surg* 1988;4:118–124.
97. Seiler T, Bende T, Woliensak J, et al: Excimer laser keratectomy for correction of astigmatism. *Am J Ophthalmol* 1988;105:117–124.
98. Lang GK, Schroeder E, Kock JW, et al: Excimer laser keratoplasty, part 2: Elliptical keratoplasty. *Ophthalmic Surg* 1989;20:342–346.
99. McDonnell PJ, Moreira H, Garbus J, et al: Photorefractive keratectomy to create toric ablation for correction of astigmatism. *Arch Ophthalmol* 1991; 109:710–713.
100. McDonnell PJ, Moreira H, Clapham TN, et al: Photorefractive keratectomy for astigmatism: Initial clinical results. *Arch Ophthalmol* 1991;109:1370–1373.
101. Waring GO: Development and classification of refractive surgical procedures. In Waring GO ed. *Refractive Keratotomy for Myopia and Astigmatism.* St Louis, Mosby–Year Book, 1992, p 145.
102. Del Pero RA, Gigstad JE, Roberts AD, et al: A

refractive and histopathologic study of excimer laser keratectomy in primates. *Am J Ophthalmol* 1990;109:419–429.

103. Lee TJ, Wan Wi-Kash RL: Keratocyte survival following a controlled rate freezing. *Invest Ophthalmol Vis Sci* 1985;26:1210–1215.

104. Gabay S, Slomovic A, Jares T: Excimer laser–processed donor corneal lenticules for lamellar keratoplasty. *Am J Ophthalmol* 1989;107:47–51.

105. Olson RJ, Kaufman HE, Rheinstrom SD: Reshaping the cat corneal anterior surface using high speed diamond fraise. *Ophthalmic Surg* 1980;11:784–786.

106. Krwawicz T: New plastic operation for correcting refractive error of aphakic eyes by changing corneal curvature: Preliminary report. *Br J Ophthalmol* 1961;45:59–63.

107. Forger W, Grewe S, Atzier U, et al: Phototherapeutic keratectomy in corneal diseases. *Refract Corneal Surg* 1993;9:585–590.

108. Sher NA, Bowers RA, Zabel RW, et al: Clinical use of 193 nm excimer laser in the treatment of corneal scars. *Arch Ophthalmol* 1991;109:491–498.

109. McDonnell PJ, Seiler T: Phototherapeutic keratectomy with excimer laser for Reis-Bucklers corneal dystrophy. *Refract Corneal Surg* 1992;8:306–310.

110. Campos M, Nielsen S, Szerenyi K, et al: Clinical follow-up of photoreactive keratectomy per treatment of corneal opacities. *Am J Ophthalmol* 1993;115:433–440.

111. Stark WJ, Chamon W, Kamp MT, et al: Clinical follow-up of 193 nm Arf excimer laser photokeratectomy. *Ophthalmology* 1992;99:805–811.

112. Brancatto R, Scialdone A, Carones F, et al: Excimer laser ablation of corneal protuberance. *J Cataract Refract Surg* 1992;18:111.

113. Fiferman RA, Forgey DR, Cook YD: Excimer laser ablation of infectious crystalline keratopathy. *Arch Ophthalmol* 1985;99:534–538.

114. Chamon W, Azar DT, Stark JW, et al: Phototherapeutic keratectomy. *Ophthalmol Clin North Am* 1993;6:399–413.

115. Azar DT: A new excimer laser technique for the correction of stabismus and diplopia. *SPIE Ophthalmic Technologies IV* 1994;2126:40–46.

CHAPTER 2

Corneal Anatomy, Transparency, and Light Scattering

Ezra L. Galler, Roger F. Steinert

FUNCTIONAL ANATOMY OF THE CORNEA

The cornea is a highly specialized avascular tissue designed to function both as part of the tough outer coat of the eye and as a clear window for the transmission and refraction of light to the interior of the eye. The cornea has four classic layers: (1) thin, stratified squamous, nonkeratinized *epithelium* on the surface; (2) thin acellular layer of collagen known as *Bowman's layer;* (3) a relatively thick connective tissue stroma with fibroblast-like cells known as *keratocytes;* and (4) an inner monolayer of low cuboidal *endothelium* with its associated basement membrane, *Descemet's membrane.*

The Epithelium

When wet, the normal epithelium has a smooth surface that acts as the major refracting interface of the eye. In humans, the corneal epithelium contains five to seven cell layers and is approximately 50 μm thick.[1] Innervation originates from the trigeminal ganglion and forms the most dense sensory nerve plexus of all epithelia, principally as protection of the exposed outer surface.[2] The nerves also have a trophic effect on the epithelium, but this function is less well understood. As elsewhere in the body, the epithelium provides a barrier to fluid loss and pathogen entrance. It is self-renewing from the mitotically active basal layer, with complete turnover in 5 to 7 days.[3] The cells have fewer than normal cytoplasmic organelles. This may contribute to the epithelial transparency.[4] The basal cells adhere to their basement membrane and to the underlying stroma through a complex series of linked microstructures called an *adhesion complex.* The epithelial cells adhere to each other through desmosomes, gap junctions, and tight junctions.[5-7] The outer or apical cell layer also has microplicae (ridge-like folds) on the surface which exhibit a filamentous glycocalyx. This interface binds the mucin layer of the tear film. The polar outer end of the mucin permits a smooth, stable aqueous tear film on the nonpolar epithelial cell membrane.[8-10]

The Stroma

The stroma is the connective tissue layer which makes up the bulk of the cornea. In humans it is approximately 500 μm thick centrally and increases to almost 1 mm in the periphery.[4,5] It is the most highly organized connective tissue in the body, which contributes to its transparency. The majority of its mass is extracellular matrix comprised mainly of collagens, together with their associated proteoglycans and soluble glycoproteins.[6] The stroma is arranged in three layers which are visibly distinct on microscopy. Bordering the epithelium is Bowman's layer (formerly incorrectly termed a *membrane*), an acellular zone 8 to 10 μm thick. The collagen fibrils are 20 to 30 nm in diameter and are woven in a random pattern into a felt-like matrix which extends across the entire cornea, tapering in thickness as it approaches the limbus.[5] There is evidence of both a stromal and an epithelial contribution to the formation of Bowman's layer.[11,12] The function of Bowman's layer is not known. Most species do not have such a layer. Primates and aves have the most highly developed Bowman's layer.[13] Whether Bowman's layer plays a critical role in development, structural integrity, or high visual acuity is unknown.

The lamellar stroma comprises the bulk of the cornea. The stroma is made up of lamellae formed from flattened bundles of collagen fibrils arranged in a parallel manner. Each bundle is 2 μm thick and 9 to 260 μm wide and runs the entire length of the corneal diameter.[14] There are 200 to 250 lamellae in the human eye.[5] In the posterior two-thirds of the stroma, the orientation of one lamella is at right angles or orthogonal to the two adjacent lamellae. The lamellae are more oblique with respect to one another in the anterior third of the lamellar stroma, and branching of lamellae has been described in this region.[14] The individual collagen fibrils are slightly larger than in Bowman's layer (27 to 35 nm in diameter), but in both regions they are considerably smaller than elsewhere in the body (e.g., the fibril diameter is up to several hundred nanometers in sclera) and are far more uniform in size.[5,13]

The lamellar stroma is secreted by the keratocytes which usually reside between lamellae.[4] The cells are very flat, with many long, slender processes extending out in all directions and making contact with the processes from nearby cells, forming an interconnected network of cells.[15] The fibrils consist primarily of collagen types I and V.[13] Additionally, immunohistochemical analysis has demostrated the presence of type XII collagen throughout the stroma. The type XII collagen may serve to link adjacent fibrils both within each lamella and between adjacent lamellae, with an observed interfibrillar spacing of 55 nm consistent with the length of the "finger" regions of the type XII molecules. The flexibility of the fingers, however, provides a limited amount of sliding between fibrils to accommodate mechanical stress.[13] The lamellar organization of the stroma provides an even distribution of tensile strength in all radial directions. The tension is also uniformly distributed throughout the stromal thickness.[17]

Descemet's Membrane

Descemet's membrane is the basement membrane of the corneal endothelium. It is secreted at the basal surface of these cells on the posterior stroma. At birth, Descemet's membrane is approximately 3 μm thick; it can reach 13 μm in thickness by late adulthood.[3,17] The anterior and oldest portion shows a banded collagen pattern.[4] There is a hexagonal lattice structure parallel to the plane of the cornea within this layer.[18] The banded layer is composed mainly of type VIII collagen, as well as fibronectin, laminin, and proteoglycans.[19] The posterior, nonbanded portion of Descemet's membrane is composed of several layers of type IV collagen-containing networks.[13] From its structure, one would expect Descemet's membrane to be an elastic structure, and this property can be demonstrated both experimentally and clinically.[16,20]

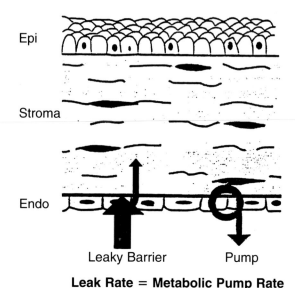

Figure 2.1. Illustration of the barrier and pump functions of the endothelium and Descemet's membrane.

The Endothelium

The corneal endothelium is a monolayer of cells at the most posterior aspect of the cornea which normally appears as a hexagonal mosaic when viewed en face. The endothelium has two major roles in overall corneal function. It allows the passage of nutrients into the stroma from the aqueous humor, and it plays the primary role in maintaining stromal dehydration and the resulting corneal transparency.[21-23] Because it is avascular, the cornea cannot receive its oxygen and nutrients from the blood. The oxygen comes mainly from the air which dissolves in the tear film and passes through the epithelium. Critical nutrients such as glucose, amino acids, and vitamins come from the aqueous humor and must pass through the endothelial cells to enter the cornea. This occurs mainly paracellularly through leaks between the cells.[24]

The other primary function of the endothelium is to maintain a relatively low level of stromal hydration. Transparency of the cornea depends on the uniformity of its component tissues and the precise pattern of their spatial arrangement, especially of the collagen bundles.[25,26] Increase in the stromal water content creates swelling and derangement of the tissue organization, resulting in loss of transparency and decreased vision. Thus, although the cornea must allow leakage of molecular nutrients from the aqueous humor, it must also serve as a barrier to the free passage of water molecules. The pump-leak hypothesis is the generally accepted explanation of how the endothelium performs these contradictory functions.[1] The focal tight junctional complexes between adjacent cells form the barrier against the aqueous fluid. Although the passage of nutrients occurs around the tight junctions, these junctions serve to retard the bulk flow of fluid into the cornea.[27-30] The excess fluid that does enter along with the nutrients must be pumped out, however. This is accomplished through activation of the Na-K-ATPase pumps located in the lateral plasma membranes of the cells. (Fig. 2.1) An unusually large number of mitochondria within the endothelial cells provide the pumps with the tremendous metabolic energy needed to drive the ion exchange.

This creates an osmotic gradient that draws water out of the cells passively into the aqueous humor.[31-33] There is also a secondary set of bicarbonate-dependent Mg-ATPase pumps which assist in the maintenance of corneal deturgescence.[34-36]

The endothelial cells are capable of normal division only during the first two trimesters of fetal life. Thereafter, cell division is limited and outpaced by natural cell loss. Hence, the number of endothelial cells decreases from that time on. At birth, the cell density is 3500 to 4000 cells/mm^2; by adulthood, it is 1400 to 2500 cells/mm^2.[37-42] However, this cell density remains more than sufficient to maintain corneal transparency in the majority of the population throughout life.

TRANSPARENCY OF THE NORMAL CORNEA

Transparency is defined as the quality of transmitting light through matter without significant scattering allowing whatever lying beyond that matter to be completely visible. This is of obvious and utmost importance for the cornea. The laws of geometric optics strictly apply only to light passing through homogeneous media. Most matter, however, is heterogeneous, either because of local variance in the density of a simple material or because of the properties inherent in composite materials. This variability within an object redirects the propagation of some of the incident light, termed *scattering*, out of

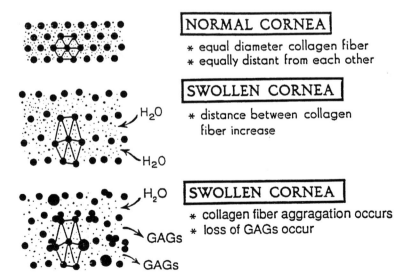

Figure 2.2. Illustration of the disruption of stromal collagen and loss of corneal transparency due to influx of water (H_2O) and loss of glycosaminoglycans (GAGs).

the incident beam. Scattering is different from *absorption,* in which some of the light energy is converted into another form, such as heat or light of another color.[43-44] The anatomic design of the cornea is thought to keep absorption of visible light to a minimum, and it is ignored in most analyses. The energy remaining in a beam of light decreases with further depth of penetration into a scattering medium and is related to the coefficient of extinction due to scattering, which is a physical characteristic of the medium. Liquids contain approximately 1000 times the number of molecules per unit volume compared with gases. One might expect 1000 times the amount of scattered light if each molecule is an independent scatterer. The observed increase of scattering in a typical pure liquid is only 50 times that of a gas, however. This is due to the fact that in a liquid, the positions of the molecules are somewhat correlated with one another, unlike those in a gas. If the correlation of molecules is perfect, as in crystalline materials, the medium is homogeneous and there should be no scattering as long as the wavelength of light is larger than the intermolecular spacing. This basic concept of light scattering is not completely true, however. Actually, the molecules of a transparent medium are all excited by the incident field and reemit scattered waves in all directions. The perfect spatial correlation of molecules creates destructive interference among the waves scattered in all but the forward direction.[45]

The corneal stroma is composed of 78% water, 1% salts, and 21% biologic macromolecules, almost 75% of which is collagen in the form of fibrils.[1] Because the refractive index of the fibrils differs from that of the ground substance in which they lie, the fibrils scatter light. If the fibrils scattered independently, one would expect 40% of an incident beam of 500-nm light to pass through the cornea. Although the cornea is not perfectly transparent (if it were, we could not view it with the slit lamp), it does transmit approximately 95% of 500-nm light. The correlation of fibril position means that the fibrils are not independent scatterers, and destructive interference eliminates much of the scattered light.[45] Maurice suggested that the fibrils are actually a perfect crystalline lattice. Since the distance between adjacent fibrils (50-60 nm) is smaller than the wavelength of light (400-700 nm), this should result in perfect transparency. Maurice attributed the minimal scattering to other stromal components such as keratocytes.[26] Electron micrographs do not show the fibrils to exist in a perfect crystalline lattice, however. This could be dismissed as artifact, but Hart and Farrell showed that the short-range order (the position of any neighboring fibril is predictable up to a distance of 200 nm from a reference fibril but is random thereafter) seen on electron micrographs could account for corneal transparency and therefore might be real.[46] Other models of short-range order have been proposed, including the correlation area model of Benedek and the hard-core cylinder model of Twersky.[47,48]

Effect of Edema on Transparency

Loss of corneal transparency due to edema has been attributed to the disruption of the regular spatial orientation and resultant loss of the destructive interference among the scattered light waves (Fig. 2.2). As the proteoglycans absorb water, the distance between the collagen fibrils is increased.[26] It is also possible that with the loss of glycosaminoglycans, fibrillar aggregation can occur, causing increased scattering of light.[49] The fibril distribution of edematous corneas seen on electron micrographs is irregular, with some areas appearing like normal cornea and other areas being completely without fibrils. The latter areas, known as *lakes*, may be the fundamental cause of the increased scattering seen with corneal edema because the linear dimensions are comparable to the wavelength of visible light.[50] Benedek's lake theory was used to show that light scattering would increase as the number and size of the lakes increased.[47]

LIGHT SCATTERING FROM POSTEXCIMER HAZE

From a clinical perspective, an increase in the degree of intraocular light scattering sufficient to veil the object of interest is perceived subjectively as disabling glare. Much interest has developed in the excimer laser as a tool both to effect refractive changes at the corneal level and to ablate superficial corneal opacities and irregularities. There is some concern because the laser most often treats the center of the cornea, and in many patients a reduction in corneal transparency seen as stromal haze develops in the ablation zone by 1 month after surgery. On average, the haze increases to a maximum at

6 to 8 weeks and then begins to subside, with little or no haze in most corneas by 6 to 12 months. The altered light reflex of the hazy area is created by the reflection and scattering of light from the atypical cellular and subcellular structures in the healing cornea.[51] This includes migratory or atypical keratocytes, microvacuoles or intralamellar inclusions, and newly synthesized atypical collagen. The haze is sometimes associated with subjective complaints of disabling glare, as well as halos and double images. We will be discussing mainly the glare, as the latter symptoms are associated with refractive discontinuities at the edge of the ablation zone rather than with light scattering.[52] Lohmann et al. proposed that the halo represents a myopic blur circle created by the differential refraction of light at the treated and untreated zones of the cornea. In a group of 58 myopes, they showed that the size of the halo was comparable for spectacle and soft contact lens wearers, as well as for patients 18 months after photorefractive keratectomy (PRK), and was much larger in hard contact lens wearers.[53]

Back Scattering

Corneal defects scatter light in all directions. Only the part scattered back toward the slit lamp (back-scattered light), plus reflected light, contribute to the opacity seen by the ophthalmologist. Clinical assessment and grading of post-excimer haze is difficult, especially as the haze evolves and becomes less homogeneous. Several investigators have used objective imaging techniques to quantify the complex back scatter and reflection from the cornea, both in normal and in post-PRK eyes.[54–56] Lohmann et al. utilized a device that differentiated between reflected and back-scattered light using polarizing filters. Both components increased to a maximum at around 2 months after excimer PRK. The scattered light was more prominent up to 3 to 4 months; the reflected light was dominant after that time. Both signals diminished to insignificant levels by 12 months in almost all patients.[57] Gartry et al. used the same device to compare post-PRK treatment with high-dose topical dexamethasone versus placebo. They found that the haze was significantly greater in the −6.0 D group compared to the −3.0 D group, but there was no difference in haze between the steroid and placebo groups.[58]

Forward Scattering

It is forward (intraocular) light scatter, however, that causes glare and visual disability (Fig. 2.3). The proportion of light scattered in the forward and backward directions depends on the geometric configuration of the corneal defect, and the two factors are not directly related to one another.[59] Therefore, it is inaccurate to estimate the amount of forward-scattered light from a measurement of the back-scattered light. However, Lohmann et al. did find a good correlation between the visual acuity loss at 5% contrast with the measure of back-scattered light and no correlation with the measure of reflected light.[57] Additionally, Esente et al. showed a correlation between the loss of high-frequency contrast sensitivity and subjectively estimated corneal haze 3 months after PRK.[60] Nevertheless, measurement of the forward-scattered light would clearly be a better way to assess the effect of the corneal haze on visual function.

Lohmann et al. measured forward-scattered light using a modification of the direct compensation method of van den Berg.[61] They also measured back-scattered light, as described earlier, and visual acuity at variable contrast levels. They also compared the post-PRK patients to myopic spectacle and contact lens wearers, as several studies had revealed decreased low-contrast visual acuity and contrast sensitivity in soft contact lens wearers. They examined a total of 35 patients. Of the 10 post-PRK patients, 5 were 3 months after surgery and 5 were at least 1 year after surgery. The van den Berg test displays a central target surrounded by an annulus of background illumination. By measuring the ratio of contrast between these two zones, a measure of intraocular scatter can be obtained. The modified version of the van den Berg test used by Lohmann et al. displayed a 1° foveally fixated center target surrounded by a 1.5° annulus, a 2.5° annulus, and finally a 5° stray light annulus. The target was set at a luminance within 20–40% of the background, and flickered between that level and the background level at 7.5 Hz. The subject had to increase or decrease the luminance of the target to minimize the flicker. This was done first without the stray light source and then with the stray light source flickering in counterphase with the target. The stray light source contributes an additional illuminance to the target caused by forward light scat-

Figure 2.3. *A:* In the normal cornea, almost all of the incident light rays are transmitted to a sharp focus on the retina. A small portion of the light (approximately 5%) is back-scattered (broken lines). The back-scattered light is seen during slit lamp examination as the normal corneal stromal haze. *B:* In a cornea with good clarity but an irregular surface, the back-scattered light remains at a normal level (broken lines) but there is increased forward scattering (dotted lines). The retinal image is degraded, with increased glare and decreased contrast sensitivity, as well as some loss of resolution (low contrast visual acuity). *C:* In a cornea with anterior scars and a smooth surface, the retinal image is degraded by decreased resolution and increased forward-light scattering (dotted lines), plus the added effect of loss of total light energy due to increased back scattering (broken lines). The ophthalmologist sees the scar as an opacity because of the back scattering. The density of scarring and the accompanying irregularity determine the total balance of decreased resolution, increased glare, and loss of contrast sensitivity.

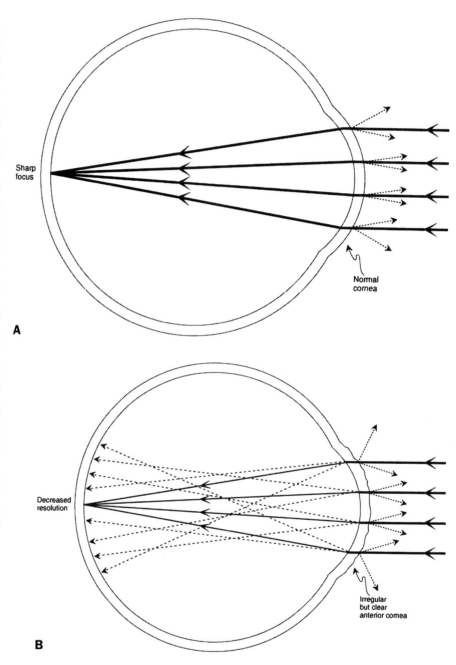

ter. Minimizing the flicker of the target yields a measure of the scattered light because the target and the stray light annulus are flickering out of phase. Soft contact lens wearers showed a decrease in 5% contrast visual acuity, and 3-month post-PRK patients were even worse, but by 12 months after PRK, low-contrast acuity was comparable to that of the spectacle and hard contact lens wearers. There was minimal difference between the three control subgroups for the measure of back-scattered light, while the 3-month PRK patients showed an increase, normalizing by 12 months. The measure of forward-scattered light, however, showed a significant increase in the soft contact lens wearers compared to those who wore spectacles or hard contact lenses, with the 3-month PRK patients were even

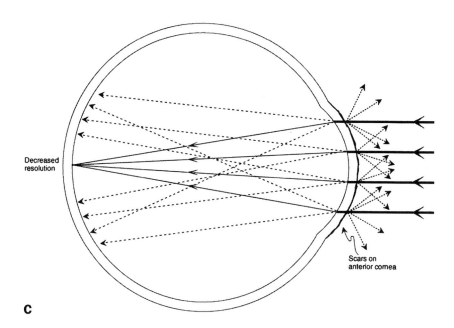

c

worse off than soft contact lens wearers but became equivalent to the spectacle group by 12 months after PRK.[51] Some authors have seen post-PRK light scattering effects last longer, and although they did not measure the forward light scatter directly, Seiler and Wollensak found a decrease in visual acuity with glare from 20/27 preoperatively to 20/31 after 1 year.[62]

Harrison et al. measured forward-scattered light in 12 myopic patients prior to and 1 month following PRK with a 6.0 mm ablation zone.[63] The patients were tested with both the stray light meter described by van den Berg and IJspeert[64] and with a computerized stray light meter developed by Lohmann et al.,[51] both based on the principal of direct compensation of van den Berg, as discussed earlier. Both of these light meters estimate forward-light scatter that would occur under normal room-lighting conditions. The patients were also tested with a mesopic increment threshold-glare paradigm which estimates the effect of forward-light scatter under nighttime illumination. This test is sensitive to light scatter from corneal defects which are not in the visual axis—for example, in eyes following radial keratotomy. The test is performed by adjusting the target (a 0.5° 2-Hz flickering flash) to threshold visibility set against a dimly illuminated background in the presence of an eccentric glare source. The corneal haze was graded subjectively on a scale from 0 to 5. Unfortunately, only some of the tests could be performed on each patient. Despite this limitation, the difference in the stray light indices before and after PRK was approximately zero in all three tests, with a larger variation from the computerized stray light meter. The authors propose a technical reason for the larger standard deviation, but it is also possible that the computerized test is more sensitive and would have yielded more significant results with a larger sample. They also measured the difference in the stray light indices between the two eyes of each patient prior to surgery, as well as between the unoperated eye without a contact lens prior to surgery and that eye with a contact lens after surgery. There was also no significant difference in these comparisons. Only three of the patients took the increment threshold test both before and after surgery, so statistical analysis was not possible. The variability among the three was small, however, and the results were consistent with those from the stray light meters. Eight additional patients took the threshold test after PRK. A comparison of the 11 post-PRK results to those of an emmetropic control group showed no significant difference. The authors concluded that the 6.0-mm treatment zone was responsible for the lack of significant forward-light scatter 1 month after surgery compared to the results with a 4-mm treatment zone,[66] although they did not do a direct comparison.

GLARE AND PHOTOTHERAPEUTIC KERATECTOMY

Several studies have looked at phototherapeutic keratectomy (PTK) specifically, and in the cases where subjective glare was a problem, it improved postoperatively. Gartry et al. treated 25 eyes with the Summit Excimed UV200. Twenty of the patients had band keratopathy and five had various other anterior stromal dystrophies. The authors used a central 4-mm treatment zone for cases of smooth band with impaired vision. They used multiple overlapping 4-mm zones for rough, painful bands. Seven of the patients had poor visual potential due to underlying pathology. Sixteen of the remaining 18 patients showed improvement in visual acuity. Seven patients were troubled by glare preoperatively, and six of them showed marked improvement, with one remaining unchanged. Of the six whose glare symptoms improved, four had improved acuity and one was unchanged. One patient had decreased acuity due to induced irregular astigmatism. This patient had been treated with the multiple overlapping zones, resulting in a faceted surface. The authors believed that central treatment of the visual axis resulted in more consistent improvement in acuity and reduction of glare.[67] Hersh et al. performed PTK on 12 eyes of 11 patients using the Summit Excimed UV200. The patients ranged from 25 to 80 years of age, and follow-up periods ranged from 1 to 4 months. Three general categories of patients were treated: (1) surface irregularities, depositions, or excrescences; (2) superficial stromal scars; and (3) recurrent erosions. Best corrected spectacle visual acuity improved in 10 of the 12 eyes, with 8 improving two or more lines. All but 1 of the 12 patients reported a subjective improvement in vision, and all 4 patients with preoperative photophobia noted an improvement as well. The authors recommended varying techniques, depending on the pathology, and used a wide range of laser pulses (15 to 1059), as well as varying the beam diameter (1.5–5 mm) from one case to another.[68] O'Brart et al. treated band keratopathy in 122 eyes of 97 patients with PTK using the Summit UV200. Two types of patients were treated: (2) smooth bands with intact epithelium and (2) rough bands with unstable epithelium. Of the 78 eyes with smooth bands, 68 had reduced vision preoperatively, and 60 improved two or more lines. Forty-eight of the smooth-band group had glare preoperatively, and 42 improved. In a subset of 17 patients with vision of 20/60 or better preoperatively, contrast sensitivity and low-contrast visual acuity were significantly improved following the laser treatment.[69] Most recently, Orndahl et al. reported on a series of 33 eyes in 31 patients with a variety of corneal dystrophies treated with PTK using both the Summit UV200 and the VISX 20/20. The treatment goal in 27 eyes was to improve vision, and 23 of them gained 2 or more lines after surgery. One patient with lattice dystrophy and three with granular dystrophy were unchanged. The goal in the other six eyes was to heal recurrent erosions and stabilize the refraction; all operations were successful. Only one patient in this study, one of the five with Schnyder's crystalline dystrophy, complained of glare preoperatively and was less bothered by it postoperatively.[70]

CONCLUSION

In summary, the cornea is an exquisitely designed organ whose microscopic and molecular structure facilitate a multiplicity of functions. The crosslinked collagen fibrils extend from limbus to limbus and, with a different orientation, from one layer to the next. This pattern creates a mechanically strong tissue to maintain the integrity of the globe. The correlation of fibril position, while not perfect, as in a crystalline structure, does create the destructive interference responsible for eliminating most forward- and back-scattered light. The optical quality of the cornea can be degraded by an irregular anterior surface or by stromal opacities which scatter light independently, as well as altering the spatial arrangement of the fibrils. Depending on the clinical setting, the excimer laser can cause the formation of a stromal haze, resulting in increased light scattering, or it can help eliminate opacities and smooth the surface to reduce light scattering and improve vision. With judicious use of the laser and with pharmacological modulation of the wound-healing response, we can utilize this technology to reduce corneal surface irregularity and anterior stromal opacity without stimulating the unwanted corneal haze formation.

REFERENCES

1. Maurice DM: The cornea and the sclera. In Davson H (ed): *The Eye*, 3rd ed. London, Academic Press, 1985, pp vol 1b pp. 1–158.
2. Rozsa AJ, Beuerman RW: Density and organization of free nerve endings in the corneal epithelium of the rabbit. *Pain* 1982;14:105.
3. Hanna C, O'Brien JE: Cell production and migration in the epithelial cell layer of the cornea. *Arch Ophthalmol* 1960;64:536.
4. Gipson IK: Anatomy of the Conjunctiva, Cornea, and Limbus. In Smolin G, Thoft RA (eds): *The Cornea*, 3rd ed. New York, Little, Brown, 1994, pp 3–24.
5. Rodrigues MM, et al: Cornea. In Jakobiec FA (ed): *Ocular Anatomy, Embryology, and Teratology*. Philadelphia, Harper & Row, 1982.
6. Gipson IK, Spurr-Michaud SJ, Tisdale AS: Anchoring fibrils form a complex network in human and rabbit cornea. *Invest Ophthalmol Vis Sci* 1987;22:212.
7. Gipson IK, Sugrue SP: Cell biology of the corneal epithelium. In Albert DM, Jakobiec FA (eds): *Principles and Practice of Ophthalmology*. Philadelphia, JB Lippincott, 1994, pp 1–37.
8. Andrews PM: Microplicae: Characteristic ridge-like folds of the plasmalemma. *J Cell Biol* 1976;68:420.
9. Gipson IK, Yankauckas M, Spurr-Michaud SJ, et al: Characteristics of a glycoprotein in the ocular surface glycocalyx. *Invest Ophthalmol Vis Sci* 1992;33:218–227.
10. Nichols BA, Chiappino ML, Dawson CR: Demonstration of the mucous layer of the tear film by electron microscopy. *Invest Ophthalmol Vis Sci* 1985;25:464.
11. Tisdale AS, Spurr-Michaud S, Rodrigues M, Hackett J, et al: Development of the anchoring structures of the epithelium in rabbit and human fetal corneas. *Invest Ophthalmol Vis Sci* 1988;29:727–736.
12. Dodson JW, Hay ED: Secretion of collagenous stroma by isolated epithelium grown in vitro. *Exp Cell Res* 1971;65:215–220.
13. Olsen BR, McCarthy MT: Molecular structure of the sclera, cornea, and vitreous body. In Albert DM, Jakobiec FA (eds): *Principles and Practice of Ophthalmology*. Philadelphia, JB Lippincott, 1994, pp 38–63.
14. Hogan MJ, Alvarado JA, Wedell JE: *Histology of the Human Eye*. Philadelphia, WB Saunders, 1971.
15. Ueda A, Nishida T, Otori T, Fujita H: Electron microscopic studies on the presence of gap junctions between corneal fibroblasts in rabbits. *Cell Tissue Res* 1987;249:473.
16. Maurice DM: Mechanics of the cornea. In Cavanagh HD (ed): *The Cornea: Transactions of the World Congress on the Cornea III*. New York, Raven Press, 1988, pp 187.
17. Johnson DH, Bourne WM, Campbell RJL: The ultrastructure of Descemet's membrane. I. Changes with age in normal corneas. *Arch Ophthalmol* 1982;100:1942.
18. Jakus MA: Studies on the cornea. II. The fine structure of Descemet's membrane. *J Biophys Biochem Cytol* 1956;2(Suppl):243.
19. Sawada H, Konomi H, Hirosawa K: Characterization of the collagen in the hexagonal lattice of Descemet's membrane: Its relation to type VIII collagen. *J Cell Biol* 1990;110:219.
20. Jue B, Maurice DM: The mechanical properties of the rabbit and human cornea. *J Biomech* 1986;10:847.
21. Waring GO III, Bourne WM, Edelhauer HF, Kenyon KR: The corneal endothelium: Normal and pathologic structure and function. *Ophthalmology* 1982;89:531.
22. Tuft SJ, Coster DJ: The corneal endothelium. *Eye* 1990;4:389.
23. Klyce SD, Beuerman RW: Structure and function of the cornea. In Kaufman HE, Banon BA, McDonald MB, Waltman SR (eds): *The Cornea*. New York, Churchill Livingstone, 1988, pp 33.
24. Joyce NC: Cell biology of the corneal endothelium. In Albert DM, Jakobiec FA (eds): *Principles and Practice of Ophthalmology*. Philadelphia, JB Lippincott, 1994, pp 17–37.
25. Mishima S: Clinical investigations on the corneal endothelium. *Am J Ophthalmol* 1982;93:1.
26. Maurice DM: The structure and transparency of the corneal stroma. *J Physiol (Lond)* 1957;136:263.
27. Iwamoto T, Smelser GK: Electron microscopy of the human corneal endothelium with reference to transport mechanisms. *Invest Ophthalmol* 1965;4:270.
28. Leuenberger PM: Lanthanum hydroxide tracer studies on rat corneal endothelium. *Exp Eye Res* 1973;15:85.
29. Kreutziger GO: Lateral membrane morphology and gap junction structure in rabbit corneal endothelium. *Exp Eye Res* 1976;23:285.
30. Montcourrier P, Hirsch M: Intercellular junctions in the developing rat corneal endothelium. *Ophthalmic Res* 1985;17:207.
31. Hodson S: The endothelial pump of the cornea. *Invest Ophthalmol Vis Sci* 1977;16:589.

32. Mayes KR, Hodson S: An in vivo demonstration of the bicarbonate ion pump of rabbit corneal endothelium. *Exp Eye Res* 1979;28:699.
33. Maurice DM: The location of the fluid pump in the cornea. *J Physiol* 1972;221:43.
34. Hodson S, Miller F: The bicarbonate ion pump in the endothelium which regulates the hydration of the rabbit cornea. *J Physiol* 1976;263:563.
35. Hull DS, Green K, Boyd M, et al: Corneal endothelium bicarbonate transport and the effect of carbonic anhydrase inhibitors on endothelial permeability and fluxes and corneal thickness. *Invest Ophthalmol Vis Sci* 1977;16:883.
36. Barfort P, Maurice DM: Electrical potential and fluid transport across the corneal endothelium. *Exp Eye Res* 1974;19:11.
37. Van Horn DL, Sendele DD, Seidman S, Buco PJ: Regenerative capacity of the corneal endothelium in rabbit and cat. *Invest Ophthalmol Vis Sci* 1977; 16:597.
38. Murphy C, Alvarado J, Juster P, et al: Prenatal and postnatal cellularity of the human corneal endothelium. A quantitative histologic study. *Invest Ophthalmol Vis Sci* 1984;25:312.
39. Bahn CF, Glassman RM, McCullem DK, et al: Postnatal development of corneal endothelium. *Invest Ophthalmol Vis Sci* 1986;27:44.
40. Laule A, Cable MK, Hoffman CE, et al: Endothelial cell population changes of human cornea during life. *Arch Ophthalmol* 1978;96:2031.
41. Landshman N, Ben-Hanam I, Assia E, et al: Relationship between morphology and functional ability of regenerated corneal endothelium. *Invest Ophthalmol Vis Sci* 1988;29: 1100.
42. Svedbergh B, Bill A: Scanning electron microscopic studies of the corneal endothelium in man and monkeys. *Acta Ophthalmol* 1972;50:321.
43. Wood RW: *Physical Optics*. New York, Macmillan, 1934.
44. van de Hulst HC: *Light Scattering by Small Particles*. New York, Dover, 1981.
45. Farrell RA: Corneal transparency. In Albert DM, Jakobiec FA (eds): *Principles and Practice of Ophthalmology*. Philadelphia, JB Lippincott, 1994, pp 64–81.
46. Hart RW, Farrell RA: Light scattering in the cornea. *J Opt Soc Am* 1969;59:766.
47. Benedek GB: The theory of transparency of the eye. *Appl Opt* 1971;10:459.
48. Twersky V: Transparency of pair-related, random distributions of small scatterers, with applications to the cornea. *J Opt Soc Am* 1975;65:524.
49. Kaye GI, et al: Further Studies of the effect of perfusion with a calcium-free medium on the rabbit cornea: Extraction of stromal components. In: JG Holley Field (ed.), *The Structure of the Eye*. New York: Elsevier Biomedical, 1982, pp. 271–278.
50. Goldman JN, Benedik GB, Dohlman CH, et al: Structural alterations affecting transparency in swollen human corneas. *Invest Opth* 7:501, 1968.
51. Lohmann CP, Fitzke F, O'Brant D, Muir MK, et al: Corneal light scattering and visual performance in myopic individuals with spectacles, contact lenses, or excimer laser photorefractive keratectomy. *Am J Ophthalmol* 1993;115:444.
52. Gartry DS, Muir MGK, Marshall J: Excimer laser photorefractive keratectomy: 18-month follow-up. *Ophthalmology* 1992;99:1209.
53. Lohmann CP, Fitzke FW, O'Brant D, Muir MK, Marshall J; Halos—a problem for all myopes? A comparison between spectacles, contact lenses, or excimer laser photorefractive keratectomy. *J Refract Corneal Surg* 1993;99(Suppl):72.
54. Smith GTH, Brown NAP, Shun-Shin GA: Light scatter from the central human cornea. *Eye* 1990;74:78.
55. Olson T: Light scattering from the human cornea. *Invest Ophthalmol Vis Sci* 1982;23:81.
56. Andrade HA, McDonald MB, Liu JC, Abdelmegeed M, et al: Evaluation of an opacity lensometer for determining corneal clarity following excimer laser photoablation. *Refract Corneal Surg* 1990;6:346.
57. Lohmann CP, Timberlake GT, Fitzke FW, Gartry DS, et al: Corneal light scattering after excimer laser photorefractive keratectomy: The objective measurements of haze. *Refract Corneal Surg* 1992;8:114.
58. Gartry DS, Muir MG, Lohman CP, Marshall J, et al: The effect of topical corticosteroids on refractive outcome and corneal haze after photorefractive keratectomy. *Arch Ophthalmol* 1992;110:944.
59. Bettelheim FA, Ali S: Light scattering of normal human lens III. Relationship between forward and back scatter of whole excised lenses. *Exp Eye Res* 1985;41:1.
60. Esente S, Passarelli N, Falco L, Passani F, et al: Contrast sensitivity under photopic conditions in photorefractive keratectomy: A preliminary study. *Refract Corneal Surg* 1993;9(Suppl):70.
61. van den Berg TJTP: Importance of pathological intraocular light scatter for visual disability. *Doc Ophthalmol* 1986;61:327.
62. Seiler T, Wollensak J: Myopic photorefractive keratectomy with the excimer laser. One-year follow-up. *Ophthalmology* 1991;98:1156.

63. Harrison JM, Tennant TB, Gwin MC, Applegate RA, et al: Forward light scatter at one month after photorefractive keratectomy. *J Refract Surg* 1995;11:83–88.
64. van den Berg TJTP, IJspeert JK: Clinical assessment of intraocular stray light. *Appl Opt* 1992;31:694.
65. Applegate RA, Trick LR, Meade DL, Hartstein J: Radial keratotomy increases the effects of disability glare: Initial results. *Ann Ophthalmol* 1987;19:293.
66. Lohmann CP, et al: "Haze" in photorefractive keratectomy: Its origins and consequences. *Lasers Light Ophthalmol* 1991;4:15.
67. Gartry D, Muir MK, Marshall J: Excimer laser treatment of corneal surface pathology: A laboratory and clinical study. *Br J Ophthalmol* 1991;75:258.
68. Hersh PS, et al: Phototherapeutic keratectomy: Strategies and results in 12 eyes. *Refract Corneal Surg* 1993;9(Suppl):90.
69. O'Brart DPS, Gartry DS, Lohmann CP, Patmore AL, et al: Treatment of band keratopathy by excimer laser phototherapeutic keratectomy: Surgical techniques and long term follow-up. *Br J Ophthalmol* 1993;77:702.
70. Orndahl M, Fagerholm P, Fitzsimmons T, Tengroth P, et al: Treatment of corneal dystrophies with excimer laser. *Acta Ophthalmol* 1994;72:235.

CHAPTER 3

Corneal Wound Healing After Excimer Keratectomy

M. Farooq Ashraf, Sandeep Jain, Mary Ann Stepp, Dimitri T. Azar

INTRODUCTION

Corneal wound healing may be the most important determinant of the final visual outcome after excimer phototherapeutic keratectomy. Excimer laser interaction with corneal tissue results in removal of the most superficial 0.20- to 0.50-μm layer with each laser pulse. After excimer laser keratectomy, the epithelium and stroma undergo a series of anatomical and physiological transformations which may continue for several months or years after surgery. Corneal wound healing involves a complex, well-regulated sequence of events characterized by activation, proliferation, and migration of epithelial corneal cells toward the wound, and synthesis and subsequent remodeling of the extracellular matrix. The extracellular matrix is composed of collagen, laminin, fibronectin, and glycosaminoglycans. Interactions between the epithelial cells and the extracellular matrix are mediated by integrins, collagen, fibronectin, other extracellular macromolecules, and growth factors. Integrins play an important role in transducing biochemical signals from the extracellular matrix to epithelial cells and keratocytes. Growth factors induce epithelial cell migration and proliferation, and control the synthesis and subsequent remodeling of extracellular matrix components. In this complex scheme, the final outcome is strongly dependent on postoperative wound healing which commonly exhibits interindividual variability.

BACKGROUND

The corneal epithelium consists of five layers of stratified squamous epithelium. The epithelial layers overlie the basement membrane, which forms a barrier between these layers and the stroma (Figure 3.1). The sheet-like basement membrane is a specialized extracellular complex of macromolecules which appears early in development and plays a role in cell differentiation and growth, selective permeability, and cell attachment. The major extracellular macromolecules of the basement membrane include type IV collagen and laminins.[1] Adhesion structures such as hemidesmosomes and anchoring filaments attach the basal epithelial cells to the stroma.[1]

The corneal stroma is an avascular structure composed of keratocytes and extracellular components. Keratocytes are differentiated mesenchymal fibroblasts that produce the extracellular matrix macromolecules, as well as the enzymes responsible for their remodeling and degradation. The extracellular matrix is composed of fibrous proteins (collagen, laminin, and fibronectin) and proteoglycans. Collagen fibrils are regularly spaced and uniform in diameter. Proteoglycans are formed when highly charged glycosaminoglycans bind to core proteins, forming large macromolecules. These proteoglycans exist in intimate association with the collagen fibrils and have been implicated as important regulators of collagen fibril spacing.[2]

Proteoglycans

The biological functions of proteoglycans are derived from the chemical characteristics of the glycosaminoglycan component of the molecule and the core protein, as well as from interactions from extracellular matrix macromolecules.[3] The stromal matrix of human cornea contains keratan sulfate proteoglycan (KSPG), choindroitin sulfate proteoglycan, heparan sulfate proteoglycan (perlecan),

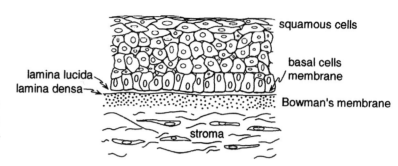

Figure 3.1. Schematic illustration of the corneal epithelium showing separation of the epithelial layers from the stroma by the basement membrane.

and dermatan sulfate proteoglycan (decorin) (Table 3.1). KSPG is the most abundant proteoglycan in the cornea, and it plays an important role in corneal transparency.

Perlecan is localized in the basement membrane, and its mRNA sequence harbors multiple domains homologous to the low density lipoprotein (LDL) receptor, laminin, neural crest adhesion molecules, and epidermal growth factors.[4] It has vital adhesive and growth regulatory functions, and binds to growth factors and protease inhibitors. Matrix cells interact with the core protein of perlecan through integrins.[5]

Decorin regulates collagen fibril formation and is a natural regulator of transforming growth factor-β (TGF-β) activity. It may prove to be clinically useful in treating fibrotic diseases caused by overproduction of TGF-β.[6] In the early stages of wound healing following full-thickness keratectomy wounds, activated keratocytes produce abnormal glycosaminoglycans and abnormally large proteoglycan filaments, especially decorin.[7-11] The abnormally large proteoglycans are most prominent 2 weeks after wounding. As healing progresses, the abnormal filaments decrease, but some persist for more than 12 months.

Laminin

Laminins are large multidomain glycoproteins located in the lamina lucida of the basal lamina. Each molecule is composed of three polypeptide chains, which form a cross-shaped structure with three short arms and one long arm.[12,13] They promote cell adhesion, growth, migration, and differentiation and play a significant role in wound healing.[14,15] Following corneal wounding, laminin first appears under the migrating cells 1 to 2 days after anterior keratectomy and becomes continuous following wound closure.

Fibronectin

Fibronectin is found in the subepithelial region and at the stromal side of Descemet's membrane in the

Table 3.1. Corneal Proteoglycans

Proteoglycan	Role in Wound Healing
Keratan sulfate (KSPG)	*Corneal transparency*
	Reduced in opaque corneal scars and wounds
	Reappears during restoration of transparency
Heparan sulfate (perlecan)	*Cell adhesive protein*
	Binds to basic FGF and protease inhibitors
	Matrix cell adhesion (through b1 and b3 integrins)
Dermatan sulfate (decorin)	*Regulates collagen fibril re-formation*
	Produced by activated keratocytes
	Increases 2 weeks after wounding
	Regulates TGF-β
Chondroitin sulfate (CS)	Increased after keratectomy wounds

unwounded cornea.[16] Its primary role is in the attachment of cells to extracellular matrix. Each chain has six domains with specific binding sites for integrins, proteoglycans, and collagen.[17] It mediates cell adhesion via interaction with integrins. The classic fibronectin receptor, α5β1 integrin, selectively binds (RGD)-containing peptides and is expressed in the wounded cornea and in cultured corneal fibroblasts. It plays an essential role in promoting cell adhesion.[16]

Cell–Matrix Interactions

The extracellular matrix (ECM) plays an active and complex role in the regulation of cells, influencing their development, migration, proliferation, shape, and metabolic functions, in addition to providing a scaffolding to stabilize the physical structure of the tissue. Matrix molecules are constantly being remodeled, degraded, and resynthesized during development. During wound healing also, there is degradation and resynthesis of matrix components. Regulating the balance of synthesis and degradation of ECM is crucial for the repair and maintenance of proper tissue architecture and clarity.

Cellular functions of ECM proteins are largely mediated by integrins, which are present on almost all cells.[19] Integrins are heterodimers having two transmembrane components, α and β subunits. Both of these subunits bind ECM proteins, and the intracellular domains interact with the actin-based cytoskeleton of the cell, influencing cellular functions. Increasing evidence indicates that integrin receptors can transduce biochemical signals from the ECM to the cell interior to modulate cell behavior (Table 3.2). Upregulation of protein kinase C activity precedes α5β1 integrin-mediated cell spreading on fibronectin.[20]

Integrins serve as receptors for laminin, vitronectin, collagen, and fibronectin.[21–23] Integrins aid in the attachment of epithelium to the basement membrane; they participate in maintaining epithelial cell shape; and, along with desmosomes, they appear to function as cell–cell adhesion molecules.[24] Integrin distribution and production are not dramatically altered during corneal epithelial cell migration over débridement wounds, except at the tip of the leading edge of migrating epithelium. However, they are available for rapid recruitment as epithelial cell migration proceeds. The major age-related difference is the loss of continuous α6 and β4 subunits along the basal surface of basal epithelial.

Masur et al. have identified integrins at the cell surface of noncultured and cultured corneal keratocytes.[25] The presence of integrins in corneal keratocytes facilitates their attachment to collagen, laminin, fibronectin, and vitronectin. When keratocytes are placed in culture, the integrin pattern changes. Integrins that bind to fibronectin are expressed, suggesting that cultured corneal keratocytes prefer fibronectin to collagen, vitronectin, or laminin as the ECM substrate.

Growth Factors

Growth factors are elements of a complex biological signaling language that provides the basis for intercellular communication. They are potent peptide regulatory factors that coordinate the proliferation, migration, and differentiation of cells and control the synthesis and remodeling of ECM. They play a crucial role in controlling ocular morphogenesis at the cellular and molecular levels.[26] In the uninjured cornea, growth factors control the balance between cell production and loss. In the injured cornea, they regulate ocular wound healing.

The normal turnover of corneal epithelium is controlled by autocrine production of growth factors such as transforming growth factor (TGF), fibroblast growth factor (FGF), and epidermal growth factor (EGF).[27–29] EGF and FGF stimulate both limbal and corneal epithelial proliferation, but platelet-derived growth factor (PDGF) stimulates only limbal epithelial proliferation.[30] Cellular differentiation is reduced during growth factor–induced mitogenic stimulation. TGF β inhibits both limbal and corneal epithelial proliferation but enhances cellular differentiation. Growth factors influence cellular migration in different dose–response patterns. Migration stimulated by basic FGF reaches a plateau, in contrast to the EGF chemotactic response, which decreases at higher concentrations.[31]

PRINCIPLES OF CORNEAL WOUND HEALING

Within the first 12–48 h after corneal wounding, epithelial cells slide down the walls of the defect and migrate over the wound.[32] As the wound is covered

Table 3.2. Integrins and Their Role in Corneal Wound Healing

Integrin Subunit Composition	Possible Function in the Healthy Cornea	Probable Function During Wound Healing
Epithelial cells		
$\alpha_6\beta_4$	Epithelial cell–substrate attachment via hemidesmisomes; progression through cell cycle	Progression through cell cycle; may mediate attachment via focal contacts on laminin
$\alpha_2\beta_1$	Cell–cell adhesion	Cell–cell adhesion
$\alpha_3\beta_1$	Cell–cell adhesion; cell–matrix adhesion to laminin	Cell–cell adhesion; cell–matrix adhesion via fibronectin or laminin
$\alpha_v\beta_5$	Cell–cell adhesion	Cell–cell adhesion; cell–matrix attachment via vitronectin or fibronectin
$\alpha_5\beta_1$	Low in level or absent in control tissues	Migration of epithelia cells and keratocytes via fibronectin (provisional matrix)
$\alpha_9\beta_1$	Cell–cell adhesion; marker for limbal stem cells and early transient amplifying cells	Cell–cell adhesion; mediates cell interaction with tenascin
Keratocytes		
$\alpha_2\beta_1$	Cell–collagen interaction; assembly of matrix	Cell–collagen interaction, assembly of matrix
$\alpha_3\beta_1$	Cell–matrix interaction; assembly of matrix	Cell–collagen interaction, assembly of matrix
$\alpha_5\beta_1$	Low level or absence of expression in control tissues	Migration of keratocytes to wound site via fibronectin

by epithelial cells, mitosis helps in the formation of multiple epithelial layers. Adhesion of the migrating epithelial cells is thought to be mediated by fibronectin. Epithelial cells have a weakened ability to adhere fully to the cornea after surgery since it may take several months for the hemidesmosomes, basement membrane, and anchoring fibrils to return to a functioning level.[33,34]

The cytoskeleton of basal cells in the leading edge of the migrating epithelium is reorganized, and the hemidesmosomes are disassembled. Focal adhesions of the migrating basal cells replace the adhesion complexes. Vinculin, α-actin, and $\alpha5\beta1$ and $\alpha3\beta1$ integrins, present in the membrane of focal adhesions, serve as provisional adhesion junctions to the newly deposited fibronectin in the wound bed.[14,35-36] In penetrating stromal wounds, repair of the anchoring fibril network resumes after reepithelialization. Patchy reformation of the basement membrane, hemidesmosomes, and anchoring fibrils appears synchronously.[14,15,35,36] Segments of the basement membrane complex become continuous by 1 to 2 months, but small areas of discontinuity and duplication may persist for longer periods.

The precise role of Bowman's layer in corneal wound healing is not well known. This layer, however, may play a key role in epithelial and stromal interactions. Hsu et al. investigated the mechanism by which anterior stromal puncture reduces the incidence of recurrent erosions. They noted an increased expression of ECM and immunolocalization of fibronectin, type IV collagen, and laminin at the site of injury.[37] They concluded that wound healing was stimulated by epithelial–stromal interactions induced by the breaching of Bowman's layer.

As early as 12 h following epithelial scrape wounds, fibronectin is deposited over the bare stromal surface in a linear fashion. It provides a temporary scaffold for corneal epithelial migration and adhesion. EGF and interleukin-6 also stimulate cell adhesion and migration by a fibronectin-dependent mechanism, possibly the increased expression of fibronectin receptors.[38] Although there have been reports that exogenously applied fibronectin promotes corneal wound healing, recent evidence suggests that exogenous fibronectin is not critical for cell adhesion and wound closure.[34,39,40]

Tenascin is a glycoprotein which has been found in skin and corneal tissues involved with wound healing. In the normal cornea, tenascin can be detected in the epithelium only. Wounding does not induce any change in epithelial tenascin. However, following anterior keratectomy, tenascin appears in the corneal stroma at the wound area only, particularly at the wound edge.[41] The precise role of tenascin in ECM interactions is still undefined.

Polymorphonuclear neutrophils (PMNs) can modulate the biosynthetic functions of corneal cells after corneal injury. Several recent studies have established a role for leukocytes and collagenases in corneal healing after alkali burns.[42,43] They may be responsible for the inhibition of epithelial proliferation. Additionally, alkali-injured corneal cells can modulate the secretion of proteins (18 kD) by PMNs.[44] Metalloproteinases produced by regenerating corneal cells after alkali burns can lead to degradation of laminin in the basement membrane zone.[45] The contribution of PMNs to ECM degradation is not fully understood. Ishizaki et al. have shown that after alkali burns, keratocytes migrate to the injured stroma, transform into myofibroblasts, and express high levels of collagen I mRNA, smooth muscle actin, and vimentin.[46] These myofibroblasts contribute to wound contraction after injury.

The corneal stroma after injury become edematous and releases inflammatory mediators.[47] The main stromal response to injury is the migration and activation of stromal keratocytes into and adjacent to the wound. These keratocytes become activated, transform into fibroblasts, migrate into the wounded area, and initiate synthesis of collagen, glycoproteins, and ECM.[48,49]

WOUND HEALING AFTER EXCIMER KERATECTOMY

Epithelium

Following excimer laser photoablation, the cornea responds by reepithelializing the surface (Figure 3.2). During corneal reepithelialization, the deposition of ECM proteins like fibronectin and tenascin precedes the regeneration of the basal lamina of the epithelium. This lends support to the idea that these proteins may serve as an intermediate scaffold for cellular migration.[50] Fibronectin occurs under epithelial cells during corneal wound healing and is believed to be important for the adherence of migrating epithelial cells to the wound surface.[51,52] Within 12 h after excimer wounding in the rabbit cornea, a band of cellular fibronectin becomes visible under the migrating epithelium over the wound surface.[50] During the next few days, the intensity of fibronectin increases and becomes more restricted to the wound edge and the anterior stroma. Nine days after wounding, only a thin layer of fibronectin is seen under the epithelium covering the wound edge. Others have shown fibronectin to be present even at 1 month in the monkey eye.[53] The pattern, time, and distribution of intrastromal fibronectin differ in excimer keratectomy and mechanical keratectomy. Tenascin appears at the wound site after 2 days, initially at the wound edge and in the anterior stroma beneath the growing epithelium, and begins to decrease after 2 weeks.[50]

Reepithelialization usually occurs in 3–5 days, with an initial thickness of three to five cells. Confocal microscopy reveals a variable cell size with superficial cells more irregular and basal epithelial cells taller than those prior to photorefractive keratectomy. The epithelium, however, continues to thicken over the next 6–8 months. Ultrastructural studies show a normal epithelium that is variably thickened. Theories of the stimulus for this epithelial hyperplasia include tear film factors and/or photoablation patterns such as the shape of the photoablated tissue (i.e., small-diameter ablations vs. larger ablations). Epithelial hyperplasia is one cause

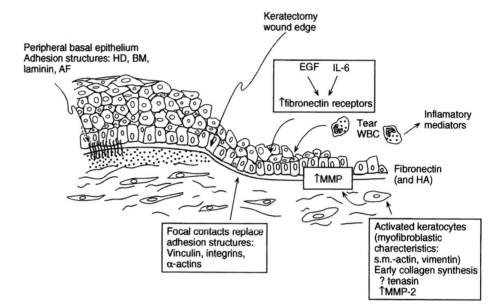

Figure 3.2. Schematic illustration of the early stages of corneal wound healing following excimer keratectomy.

of the refractive changes seen in the first 3–6 months and is a factor in the regression of the refractive effect that can occur after photorefractive keratectomy (Figure 3.3).

There is variability in the basement membrane, at times there are multiple focal discontinuities, and in some areas no basement membrane is present.[54] In a few of these areas the basal epithelial layer projects into the superficial stroma, presumably to fill in minute stromal defects.[55] The basement membrane, however, appears to be histologically normal.

In general, epithelial migration to resurface the stroma occurs rapidly. The stability of this epithelium requires the reestablishment of tight adhesions of the epithelium to the underlying stroma. The anchoring fibrils, basement membrane, and hemidesmosomes are important structures in establishing these epithelial adhesions. SundarRaj et al. showed that immediately after photoablation in monkey eyes, the basement membrane zone containing the anchoring fibrils is removed and there is no immunostaining for type VII collagen, which is present in the nonablated basement membrane.[56] Re-formation of the anchoring fibrils in the healing corneas was evaluated by the distribution of type VII collagen, the major component of anchoring fibrils. By day 7, the ablated zone was reepithelialized and began to secrete type VII collagen in a punctate linear pattern. Linear deposition of fibronectin was detectable in some regions; type III collagen was not detectable. At 3 weeks, type VII collagen staining appeared thicker than normal corneas and exhibited a segmented pattern of distribution. This change from punctate to more segmented discontinuities from day 7 to later stages of healing suggested that a continuous degradation process probably accompanied the regeneration of the anchoring fibril zone. Although recurrent erosions were not seen clinically, the epithelium overlying the basement membrane zone, which was devoid of anchoring fibrils, appeared to be less firmly attached.[56]

On the basis of rabbit models, it has been accepted that adhesion structures between corneal epithelium and stroma are restored approximately 6–8 weeks after injury. Morphometric characteristics of human epithelial adhesion structures after excimer photokeractectomy showed abnormalities in the adhesion complexes 6–15 months after keratectomy, and the reappearance of anchoring fibrils is directly related to the duration of wound healing.[57] A higher percentage of basal epithelial cells had normal anchoring fibrils at 15 months (35%) than at 6 months (8%). Other factors that may influ-

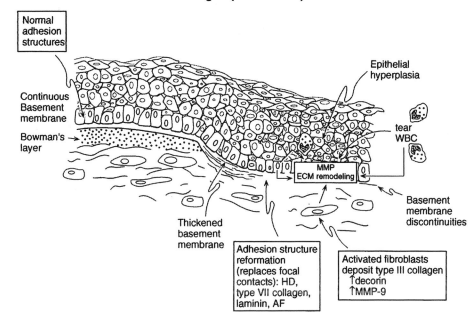

Figure 3.3. Schematic illustration of the late stages of corneal wound healing following excimer keratectomy.

ence epithelial adhesion after excimer keratectomy include the age of the patient, the depth of keratectomy, and the nature of the underlying condition.[57]

Marshall and colleagues noted the formation of a pseudomembrane after excimer keratotomies.[58] A few epithelial cells immediately adjacent to the ablated area had a pale-staining cytoplasm. These pale-staining cells have no cell membrane along the border adjacent to the ablated area but are cleaved by the photoablation process. The severed edge was bounded by an electron-dense condensation less than 100 nm thick which seemed to maintain the integrity of the cell, suggesting a membrane-like function.

Although the cause of the pseudomembrane is unknown, it may be formed from uncoupled organic double bonds created during photoablation or possibly is due to a thermal effect. The differences in pseudomembrane appearance and thickness under different hydration conditions suggest a possible thermal mechanism, but even if this is so, there is no conductive damage adjacent to the pseudomembrane.

Early photorefractive keratectomy studies used nitrogen gas blowing to maintain relative dehydration of the cornea, since photoablation is strongly dependent on the level of tissue hydration. When nitrogen gas blowing is used, the pseudomembranes appear thicker, with frequent surface discontinuities and detached fragments compared to the condition achieved with humidified blowing.[59] Corneal dehydration seems to have a detrimental effect on the pseudomembrane, while hydration at physiological levels keeps it thin and uniform.

The structural integrity of the pseudomembranes allows it to wrinkle during critical point drying prior to scanning electron microscopy.[60] It maintains the integrity of sectioned cells, at least temporarily, and its smooth, uniform surface may appear to serve as a template for reepithelialization during healing. Finally, the pseudomembrane serves as a barrier to water transportation and prevents significant corneal swelling after photorefractive keratectomy.[60]

Bowman's Layer

Studies of the cornea after phototherapeutic keratectomy show a smooth, sharp excision of Bowman's layer at the ablation site and the absence of Bowman's membrane at the center of the ablation zone. Bowman's membrane does not regenerate after its removal, suggesting that the absence of this layer may not be entirely necessary for corneal clarity, as evidenced by the lack of corneal erosions and the presence of normal clarity after phototherapeutic keratectomy. However, destruction of Bowman's layer by infection or by other surgical procedures has resulted in corneal scarring.

Stroma

Stromal wound healing is dependent on a strong interaction between the epithelium and the stromal keratocytes. The epithelium releases cytokines to activate the keratocytes and to initiate stromal proliferation. Immediately after excimer keratectomy there is mild stromal edema, and stromal wound healing occurs as inflammatory cells invade the stroma from the tear film. PMNs can be seen after deepithelialization alone, but they increase significantly after photorefractive keratectomy. Chew et al., however, using the confocal microscope on humans after photorefractive keratectomy, did not find any evidence of inflammatory cells in the stroma.[61] Fibrin and fibronectin, from the tear film, may act as a scaffold for the adhesion of neutrophils, macrophages, and lymphocytes. Tear film plasmin levels are also markedly elevated during the postoperative period and signify involvement of the plasminogen activator system, which facilitates degradation, removal, and repair of damaged collagen and ECM. Hyaluronic acid (HA) is not normally found in the corneal stroma, but its expression following excimer wounds has been reported.[62] It may represent a nonspecific corneal tissue response to injury. Exogenous HA promotes corneal reepithelialization. Inoue and Katakami have compared the effect of HA on epithelial cell proliferation in cultured corneas with that of EGF and fibronectin.[63] HA stimulated epithelial proliferation more than EGF did, and the results of the study supported the possible existence of HA receptors in corneal epithelial cells, by which HA stimulates cell proliferation.

The confocal microscope reveals that the subepithelial stroma undergoes dense fibroplasia and scarring, with several large, clear lacunae 50–100 μm in diameter.[64] These areas lacking cells or collagen were coincident with subepithelial haze seen with the slit lamp. Pockets of intercellular edema in the superficial stroma was also seen; these were not seen in routine histological slides. In rabbits immediately following photorefractive keratectomy, keratocytes in the anterior 40 μm of the stromal layers disappear and there is no secretion of extracellular material at that time.[65] Repopulation begins once the epithelium has covered the defect. The keratocytes not only repopulate but also increase in density after excimer keratectomy.[65,66] By the third postoperative day they triple in number. This is further evidence of a strong stromal–epithelial interaction.[65]

The keratocytes not only proliferate but also undergo fibroblastic transformation to help synthesize new collagen and proteoglycan matrix.[63] Immunostaining in the rabbit cornea immediately following ablation reveals immunolocalization of type IV collagen, proteoglycans, fibronectin, and laminin in a pattern similar to that of normal wound healing. On day 7, a narrow band of type VII collagen is evident along the basement membrane zone. This stained zone is narrower and punctate compared to normal cornea, which shows a continuous linear band. Fibronectin is detectable in some subepithelial regions. There is no immunostaining of type III collagen.[55] The anterior stromal ablation zones in monkeys also reveal changing segmental discontinuities even after 18 months, suggesting a continuous degradation process.

Following excimer keratectomy the refractive error is often unstable, especially during the first few weeks. To obtain a stable refractive effect, the ablated tissue ideally should not be replaced. The change in refractive effect is often due to epithelial hyperplasia, as previously mentioned, and to the regeneration of removed stromal tissue. SundarRaj et al. estimated that 10–20 μm of the centrally ablated tissue was newly formed and that type III collagen, which is not found in unwounded monkey or human corneas, increased during healing up to 18 months after treatment.[49]

Corneal transparency is dependent on a uniform cross-sectional diameter and on adequate spacing between the collagen fibrils to form a regular lattice network. Disorganized collagen is one factor responsible for the early light scatter and corneal haze seen postoperatively (Figure 3.4). Widely spaced collagen has been documented in rabbit corneal stroma following a variety of injuries.[56] Initially, irregular collagen with irregular spacing is deposited. This can be appreciated with the use of the fluorescent dye dichlorotriazinyl aminofluorescein (DTAF), which fluoresces only normal corneal stroma. Immediately after excimer ablation there is

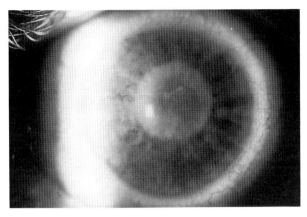
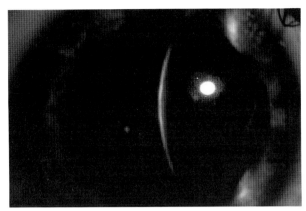
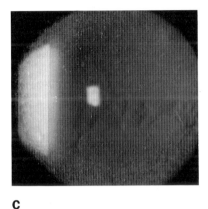

Figure 3.4. A 29-year-old man with lattice dystrophy presenting with 20/200 vision who underwent phototherapeutic keratectomy in the right eye. A mean haze grade of 1+ was noted 3 months postoperatively; visual acuity at that time was 20/100. *A:* Clinical appearance of the cornea 6 months postoperatively, with visual acuity of 20/70. *B, C:* At 30 months postoperatively, by direct illumination and retroillumination, respectively. Visual acuity improved to 20/40.

uniform fluorescence of the dye. As stromal healing begins and new collagen is produced by the keratocytes, a patchy staining pattern is seen along the border between newly formed and old collagen.

Components of the ECM may also contribute to the corneal haze seen after excimer keratectomy, since this is also altered. Glycosaminoglycans are extremely hydrophilic and play a vital role in the regulation of tissue hydration, collagen fibril diameter, and spacing. These are key factors in corneal transparency. After superficial keratectomy wounds, proteoglycans increase in size and contain predominantly chrondroitin and dermatan sulfate, with negligible levels of normally sulfated keratan sulfate.[2] The highly sulfated dermatan and chondroitin sulfate may contribute to the increased stromal water content and the corneal haze seen after excimer keratectomy. With time stromal remodeling occurs, as evidenced in the study of SundarRaj et al., in which newly formed collagen III slowly decreased and keratan sulfate proteoglycan levels increased over a follow-up period of 18 months.[49] The resorption and remodeling of collagen is a key component in corneal wound healing. Resorption is initiated by proteolytic enzymes from PMNs, macrophages, epithelial cells, and keratocytes. The degradation of collagen is mediated by metalloproteinases, including collagenase and gelatinase. These metalloproteinases are increased after corneal injury. However, with continued resorption and remodeling, fibroblasts and irregular collagen gradually disappear and reasonably normal lamellar stromal structure reappears, with less stromal scarring.

Descemet's Membrane/Endothelium

In rabbits an unusual fibrillar response in the middle to anterior one-third of Descemet's membrane and a slight enlargement of the intracellular spaces, increased visibility of the endoplasmic reticulum, and the appearance of granular material at the basal side and in intracellular spaces of endothelial cells have been observed following photorefractive keratectomy.[65] Immunofluorescence staining shows increased amounts of fibronectin, proteoglycans, and laminin in these areas, suggesting that this fibrogranular material may have been secreted by stimulated endothelial cells. Acoustic shock waves from excimer photoablation may stimulate the endothelial cells to secrete abnormal material. Zabel et al. observed endothelial damage in the posterior stroma of bovine corneas after 85% of the stroma had been ablated by the excimer laser.[67]

No significant abnormalities in Descemet's membrane or the endothelium have been reported in humans, with the exception of one patient following phototherapeutic keratectomy for superficial corneal opacities, for which a penetrating keratoplasty was eventually performed. Discrete electron-dense, collagen-like fibers were observed parallel to the posterior surface of Descemet's membrane. These were seen only under the photoablated area. No clinical abnormalities in Descemet's membrane following photorefractive keratectomy in humans have been reported. Chew et al., using the confocal microscope after photorefractive keratectomy in humans, noted that the posterior stroma, Descemet's membrane, and endothelium under the ablation were identical to those of the untreated cornea.[61]

THE ROLE OF GROWTH FACTORS IN CORNEAL WOUND HEALING

Epidermal Growth Factor

EGF stimulates corneal epithelial proliferation and migration and proliferation of stromal fibroblasts, increases collagen and fibronectin synthesis, and increases fibronectin receptor activity.[64,68] EGF and TGF-α are members of a family of single-chain polypeptide growth factors that bind to a transmembrane tyrosine kinase receptor to stimulate protein synthesis and cellular proliferation. The EGF-induced mitogenic effect is characterized by its dose-dependent downregulation.[30] Recently, EGF has been used to promote corneal reepithelialization following alkali burns, epikeratoplasty, traumatic corneal ulcers, and herpetic ulcers.[69-73] It increases the tensile strength of corneal incisions.[74] To prevent receptor downregulation during treatment, Sheardown et al. have used a controlled-release system to deliver a continuous optimal dose of EGF for a prolonged period, in preference to multiple topical eyedrops of higher concentration.[75]

Transforming Growth Factor

TGF-α is a member of the EGF family with approximately 40% homology to EGF, and it binds to the EGF receptor.[76] It is essential for normal ocular morphogenesis and has been localized in the human corneal epithelium.[77] Autocrine production of TGF-α may control the normal turnover of corneal epithelium.[78]

TGF-β is a prototypical multifunctional growth factor. The three known mammalian TGF-β isoforms (TGF-β1-3) are homo- or heterodimers with overlapping biological functions mediated by three distinct cell surface receptors (TGF-β receptor types 1-3). TGF-β2 has been localized in the superficial limbal epithelial cells of the human cornea.[79] This location is consistent with its possible role in the transdifferentiation of conjunctival to corneal epithelium. It has been localized in corneal fibroblasts following epithelial wounding. In vitro studies on human corneal stromal fibroblasts reveal that TGF-β stimulates proliferation and motility of stromal fibroblasts and increases collagen and fibronectin synthesis.[64,80] In addition to stimulating the synthesis of ECM, TGF-β suppresses matrix degradation by collagenases. It decreases the synthesis of proteases that degrade matrix proteins (matrix metalloproteinases) and increases the synthesis of protease inhibitors that block the activity of such proteases. In addition, it increases the interaction of cells with the ECM, possibly by increasing the expression of integrin receptors. TGF-β1 causes a dose-related inhibition of epithelial cell proliferation.[30] By itself it does not affect cell adhesion or migration, but it inhibits the stimulatory effects of EGF, suggesting that it serves as a modulator of EGF.[81] Phillips et al. have examined the TGF-β-induced initiation and pattern of corneal angiogenesis.[82,83] TGF-β stimulates angiogenesis indirectly by recruiting inflammatory cells. Preventing the inflammation with methylprednisolone blocks subsequent angiogenesis. TGF-β may be useful in promoting corneal stromal wound healing, as in the treatment of a corneal ulcer.

Fibroblast Growth Factors

FGF comprise a family of at last five structurally related proteins, of which acidic and basic FGF are the prototypes. Both types of FGF affect cell proliferation and differentiation.[30,84] Acidic FGF has been localized in the corneal epithelium.[85] It is overexpressed during active epithelial migration.[86] Basic FGF is present mainly in the ocular basement membranes, complexed with heparan sulfate proteoglycan.[85,87] In the intact cornea, it has been localized in the epithelial cells; only small amounts are present in Bowman's layer.[29,88] Epithelial cells do not actively secrete basic FGF. Injury-related passive release of intracellular basic FGF is the predominant method by which it is deposited in Bowman's layer.[87] FGF is endowed with a variety of biological activities, the most striking of which are related to wound healing. FGF stimulates epithelial cell proliferation and promotes stromal wound healing, stimulating the fibroblasts to proliferate and to synthesize ECM components.[90,91] Topical application of basic FGF promotes corneal epithelial and stromal healing without causing morphologically adverse reactions or intraocular and systemic penetration.[92] FGF is highly angiogenic and may be responsible for corneal vascularization.[41]

Platelet-Derived Growth Factor

PDGF is a dimeric molecule composed of A and/or B polypeptide chains (PDGF-AA/AB/BB). It is a major mitogen and chemoattractant. The presence of PDGF receptors in human corneal epithelium, fibroblasts, and endothelium, as well as the mitogenic effects of PGDF on corneal cells, suggest that it may play a role in corneal wound healing.[93]

WOUND HEALING THERAPEUTIC STRATEGIES

In theory, correction of refractive errors using the excimer laser should be ideal since the laser produces precise, reproducible stromal ablation, thus changing the refractive power of the cornea. In the complex biological model, however, refractive change is strongly dependent on postoperative wound healing. In corneal wound healing after photoablation, the epithelium must retain normal thickness, with no recurrent erosions, and the subepithelial stroma should retain a normal architecture, with no regression or scarring. The two main problems after phototherapeutic keratectomies are subepithelial haze and regression of the refractive effect secondary to epithelial hyperplasia or stromal collagen remodeling. With these factors in mind, numerous investigators have tried to modify these changes through either pharmacological means or different phototherapeutic parameters.

Corticosteroids and antimetabolites, mitomycin C, and 5-fluorouracil have been used postoperatively to inhibit wound healing and to prevent fibrosis. Corticosteroids reduce corneal scarring in animal studies of excimer keratectomies, and steroids have been used topically in most postoperative regimens in human clinical trials. The exact mechanism of action is unknown; however, reduced DNA synthesis and collagen deposition in activated fibroblasts, and prevention of the recruitment of macrophages and lymphocytes, have been implicated. Despite their beneficial qualities, however, both corticosteroids and antimetabolites have significant potential side effects.

Talamo et al. showed that in rabbits treated with topical mitomycin C and topical steroids after excimer keratectomy, subepithelial collagen synthesis was inhibited and there was less stromal haze.[94] An additive effect with concurrent use of both mitomycin C and steroids was also noted. There was, however, no difference in efficacy between the two drugs.

In a similar study using rabbits, Bergman and Spigelman attempted to demonstrate the effect of topical 5-fluorouracil, corticosteroids, and heparin on corneal wound healing following photoablation.[95] Corneal stromal haze developed in all eyes. Following 2 weeks of treatment, there was a statistically significant decrease in stromal haze in the treated groups compared to the nontreated groups. There was no significant difference between the various treated groups. At 6 weeks, however, there was no significant difference in stromal haze between all of the groups. The authors concluded that the pharmacological agents provided only a transient benefit in reducing corneal haze.

O'Brart et al. conducted a prospective randomized study on the effects of topical corticosteroids and plasmin inhibitors after photokeratectomies in humans, with a follow-up period of 12 months.[96] Patients were given topical fluorometholone 0.1%, aprotinin 40 IU/ml (a plasmin inhibitor), or no treatment. Refractive outcomes revealed a hyperopic shift in the steroid-treated groups, especially in the first few weeks after the procedure. This hyperopic shift became statistically insignificant by the sixth postoperative month. Four patients (14%), however, did retain a hyperopic shift at 12 months. There was no refractive difference between the aprotinin-treated and control groups. Stromal haze reached a maximum at 3 months for the −3.00 D corrected group and at 6 months for the −6.00 D corrected group; thereafter, haze declined in both groups. The results showed that there was no significant difference in haze in all three groups at any stage. The authors concluded that long-term use of corticosteroids should not be instituted routinely after photorefractive keratectomy, especially in light of the potential side effects associated with these agents. They also concluded that the plasmin inhibitor aprotinin had no beneficial effect on haze or the refractive outcome.

Prevention of corneal scarring and fibrosis using native ECM factors such as EGF and TGF-β is currently being investigated. These may prove to better alternatives to modulate corneal wound healing, thereby preventing corneal scarring and refractive change after excimer keratectomies.

REFERENCES

1. Yurchenco PD, Schittny JC: Molecular architecture of the basement membranes. *FASEB J* 1990;4:1577–1590.
2. Hassell JR, Cintron C, Kublin C, Newsome DA: Proteoglycan changes during restoration of transparency in corneal scars. *Arch Biochem Biophys* 1983;222:362–369.
3. Yanagishita M: Function of proteoglycans in the extracellular matrix. *Acta Pathol Jpn* 1993;43:283–293.
4. Noonan DM, Fulle A, Valene P, et al: The complete sequence of perlecan, a basement membrane heparin sulfate proteoglycan, reveals extensive similarity with laminin A chain, low density lipoprotein receptor, and the neural cell adhesion molecule. *J Biol Chem* 1991;34:22939–22947.
5. Hayashi K, Madri JA, Yuchenco PD: Endothelial cells interact with the core protein of basement membrane proteoglycan through beta 1 and beta 3 integrins: An adhesion modulated by glycosaminoglycans. *J Cell Biol* 1992;119:945–959.
6. Noble NA, Harper JR, Border WA: In vivo interactions of TGF-beta and extracellular matrix. *Prog Growth Factor Res* 4(4):369–82, 1992;4:369–382.
7. Cintron C, Covington HI, Kublin CL: Morphological analysis of proteoglycans in rabbit corneal scars. *Invest Ophthalmol Vis Sci* 1990;31:1789–1798.
8. Cintron C, Gregory JD, Damle SP, Kublin CL: Biochemical analysis of proteoglycans in rabbit corneal scars. *Invest Ophthalmol Vis Sci* 1990;31:1975–1981.
9. Funderburgh JL, Chandler JW: Proteoglycan of rabbit corneas with nonperforating wounds. *Invest Ophthalmol Vis Sci* 1989;30:435–442.
10. Rawe IM, Tuft SJ, Meek KM: Proteoglycan and collagen morphology in superficially scarred rabbit cornea. *Histochem J* 1992;24:11–18.
11. Rawe IM, Zabel RW, Tuft SJ, Chen V, Meek KM: A morphological study of rabbit corneas after laser keratectomy. *Eye* 1992;6:637–642.
12. Kleinman HK, Weeks BS, Schnaper HW, Kibbey MC, Yamamura K, Grant DS: The laminins: A family of basement membrane glycoproteins important in cells differentiation and tumor metastasis. *Vitam Horm* 1993;47:161–186.
13. Engel J: Laminins and other strange proteins. *Biochemistry* 1992;31:10643–10651.
14. Gipson IK, Spurr-Michaud SJ, Tisdale AS, et al: Reassembly of the anchoring structures of the corneal epithelium during wound repair in the rabbit. *Invest Ophthalmol Vis Sci* 1989;30:425–434.
15. Stock EL, Kurpakus MA, Sambol B, Jones JC: Adhesion complex formation after small keratectomy wounds in the cornea. *Invest Ophthalmol Vis Sci* 1992;33:304–313.
16. Tervo T, Sulonen J, Valtones S, Vannas A, Virtanen I: Distribution of fibronectin in human and rabbit corneas. *Exp Eye Res* 1986;42:399–406.
17. Gipson IK, Watanabe H, Zieske JD: Corneal wound healing and fibronectin. *Int Ophthalmol Clin* 1993;33:149–163.
18. Koivunen E, Gay DA, Ruoslahti E: Selection of peptides binding to the alpha 5 beta 1 integrin from phage display library. *J Biol Chem* 1993;268:20205–20210.
19. Virtanen I, Tervo K, Korhonen M, Paallysaho T, Tervo T: Integrins as receptors for extracellular matrix proteins in human cornea. *Acta Ophthalmol* 1992;70:18–21.
20. Vuori K, Ruoslahti E: Activation of protein kinase C precedes alpha 5 beta 1 integrin-mediated cell spreading on fibronectin. *J Biol Chem* 1993;268:21459–21462.

21. Lauweryns B, van den Oord JJ, Volpes R, Foets B, Missotten L: Distribution of very late activation integrins in the human cornea. An immunohistochemical study using monoclonal antibodies. *Invest Ophthalmol Vis Sci* 1991;32:2079–2085.
22. Hynes RO: Integrins: Versatility, modulation, and signalling in cell adhesion. *Cell* 1992;69:11–25.
23. Svennevik E, Linser PJ: The inhibitory effects of integrin antibodies and the RGD tripeptide on early eye development. *Invest Ophthalmol Vis Sci* 1993;34:1774–1784.
24. Stepp MA, Spurr-Michaud S, Gibson IK: Integrins in the wounded and unwounded stratified squamous epithelium of the cornea. *Invest Ophthalmol Vis Sci* 1993;34:1829–1844.
25. Masur SK, Cheung JK, Antohi S: Identification of integrins in cultured corneal fibroblasts and in isolated keratocytes. *Invest Ophthalmol Vis Sci* 1993;34:2690–2698.
26. Tripathi BJ, Tripathi RC, Livingston AM, Borisuth NS: The role of growth factors in the embryogenesis and differentiation of the eye. *Am J Anat* 1991;192:442–471.
27. Wilson SE, He YG, Lloyd SA: EGF, EGF receptor, basic FGF, TGFb1, and IL-1 alpha messenger RNA production in the human corneal epithelial cells and stromal fibroblasts. *Invest Ophthalmol Vis Sci* 1992;33:1756–1765.
28. Wilson SE, Lloyd SA, He YG: EGF, basic FGF, and TGF beta-1 messenger RNA production in rabbit corneal epithelial cells. *Invest Ophthalmol Vis Sci* 1992;33:1987–1995.
29. Wilson SE, Lloyd SA, He YG: Fibroblast growth factor-1 receptor messenger RNA expression in corneal cells. *Cornea* 1993;12:249–254.
30. Kruse FE, Tseng SC: Growth factors modulate clonal growth and differentiation of cultured rabbit limbal and corneal epithelium. *Invest Ophthalmol Vis Sci* 1993;34:1963–1976.
31. Grant MB, Khaw PT, Schultz GS, Adams JL, Shimizu RW: Effects of epidermal growth factor, fibroblast growth factor, and transforming growth factor-beta on corneal cell chemotaxis. *Invest Ophthalmol Vis Sci* 1992;33:3292–3301.
32. Fountain TR, de la Cruz Z, Green WR, Stark WJ, Azar DT: Reassembly of corneal epithelial adhesion structures after excimer laser keratectomy in humans. *Arch Ophthalmol* 1994;112:967–972.
33. Berman M, Manseau E, Law M, Aiken D: Ulceration is correlated with degradation of fibrin and fibronectin at the corneal surface. *Invest Ophthalmol Vis Sci* 1983;24:1358–1366.
34. Wu CS, Stark WJ, Green WR: Corneal wound healing after 193 nm excimer laser keratectomy. *Arch Ophthalmol* 1991;109:1426–1432.
35. Gipson IK, Spurr-Michaud SJ, Tisdale AS, et al: Hemidesmosomes and anchoring fibril collagen appear synchronously during development and wound healing. *Dev Biol* 1988;126:253–262.
36. Gipson IK, Spurr-Michaud SJ, Tisdale AS, et al: Redistribution of hemidesmosome components during migration of the corneal epithelium. *Invest Ophthalmol Vis Sci* 1991;32:1163A.
37. Hsu JKW, Rubinfeld RS, Barry P, Jester JV: Anterior stromal puncture—Immunohistochemical studies in human corneas. *Arch Ophthalmol* 1993;111:1131–1137.
38. Nishida T, Nakamura M, Mishima H, Otori T: Interleukin 6 promotes epithelial migration by a fibronectin-dependent mechanism. *J Cell Physiol* 1992;153:1–5.
39. Kim KS, Oh JS, Kim IS, Jo JS: Clinical efficacy of topical homologus fibronectin in persistent corneal epithelial disorders. *Korean J Ophthalmol* 1992;6:12–18.
40. Nishida T, Nakagawa S, Nishibayashi C, Tanaka H, Manabe R: Fibronectin enhancement of corneal epithelial wound healing of rabbits in vivo. *Arch Ophthalmol* 1984;102:455–456.
41. Tervo K, Tervo T, van Setten GB, Tarkkanen A, Virtanen I: Demonstration of tenascin-like immunoreactivity in rabbit corneal wounds. *Acta Ophthalmol* 1989;67:347–350.
42. Chayakul V, Reim M: The enzymatic activities in the alkali-burned rabbit cornea. *Graefes Arch Clin Exp Ophthalmol* 1982;218:145–148.
43. Pahlitzsch T, Sinha P: The alkali-burned cornea: Electron microscopical, enzyme histochemical, and biochemical observations. *Graefes Arch Clin Exp Ophthalmol* 1985;223:298–306.
44. Kao WWY, Zhu G, Kao CWC: Effects of polymorphonuclear neutrophils on protein synthesis by alkali-injured rabbit corneas—A preliminary study. *Cornea* 1993;12:522–531.
45. Saika S, Kobata S, Hashizume N, Okada Y, Yamanaka O: Epithelial basement membrane in alkali-burned corneas in rats—an immunohistochemical study. *Cornea* 1983;12:383–390.
46. Ishizaki M, Zhu G, Haseba T, Shaefer SS, Kao WWY: Expression of collagen-I, smooth muscle alpha-actin, and vimentin during the healing of alkali-burned and lacerated corneas. *Invest Ophthalmol Vis Sci* 1993;12:3320–3328.

47. Eiferman RA, Schultz GS, Nordquist RE, Waring GO: Corneal wound healing and its pharmacologic modification after refractive keratotomy. In Waring GO (ed): *Refractive Keratotomy for Myopia and Astigmatism.* St Louis, Mosby Year Book, 1992, pp 749–779.
48. Kitano S, Goldman JN: Cytologic and histochemical changes in corneal wound repair. *Arch Ophthalmol* 1966;76:345–354.
49. SundarRaj N, Geiss MJ, Fantes F, et al: Healing of excimer laser ablated monkey corneas, an immunohistochemical evaluation. *Arch Ophthalmol* 1990; 108:1604–1610.
50. Van Setten GB, Koch JW, Tervo K, et al: Expression of tenascin and fibronectin in the rabbit cornea after excimer laser surgery. *Graefes Arch Clin Exp Ophthalmol* 1982;230:178–183.
51. Fujikawa LS, Foster CS, Harrist TJ, Lanigan JM, Colvin RB: Fibronectin in healing rabbit corneal wounds. *Lab Invest* 1981;45:120–129.
52. Fujikawa LS, Foster CS, Gipson IK, Colvin RB: Basement membrane components in healing rabbit corneal epithelial wounds: Immunofluorescence and ultrastructural studies. *J Cell Biol* 1984;98:128–138.
53. Malley DS, Steinert RF, Puliafito CA, Dobi ET: Immunofluorescence study of corneal wound healing after excimer laser anterior keratectomy in the monkey eye. *Arch Ophthalmol* 1990;108:1316–1322.
54. Hanna KD, Pouliquen YM, Savoldelli ME, et al: Corneal wound healing in monkeys 18 months after excimer laser photorefractive keratectomy. *Refract Corneal Surg* 1990;6:340–345.
55. Wu WCS, Stark WJ, Green WR: Corneal wound healing after 193 nm excimer laser keratectomy. *Arch Ophthalmol* 1991;109:1426–1432.
56. SundarRaj N, Geiss MJ, Fantes F: Healing of excimer laser ablated monkey corneas, an immunohistochemical evaluation. *Arch Ophthalmol* 1990; 108:1604–1610.
57. Fountain TR, de la Cruz Z, Green WR, Stark WJ, Azar DT: Reassembly of corneal epithelial adhesion structures after excimer laser keratectomy in humans. *Arch Ophthalmol* 1994;112:967–972.
58. Marshall J, Trokel S, Rothery S, et al: Long term healing of the central cornea after PRK using an excimer laser. *Ophthalmology* 1988;95:1411–1421.
59. Campos M, Cuevas K, Garbus J, et al: Corneal wound healing after excimer laser ablation: Effects of nitrogen gas blower. *Ophthalmology* 1992;99: 893–897.
60. Gordon M, Brint SF, Durrie DS, et al: Photorefractive keratectomy at 193 nm using an erodible mask. In Parel JM (ed): *Ophthalmic Technologies II.* Bellingham, Wash, SPIE, 1992.
61. Chew SJ, Beurman RW, Kaufman HE, McDonald MB: In vivo confocal microscopy of corneal wound healing after excimer laser photorefractive keratectomy. *CLAO J* 1995;21:273–280.
62. Fitzsimmons TD, Fagerholm P, Harfstrand A, Schenhalm M: Hyaluronic acid in the rabbit cornea after excimer laser superficial keratectomy. *Invest Ophthalmol Vis Sci* 1992;33:3011–3016.
63. Inoue M, Katakami C: The effects of hyaluronic acid on corneal epithelial cell proliferation. *Invest Ophthalmol Vis Sci* 1993;34:2313–2315.
64. Ohji M, Sundar-Raj N, Thoft RA: Transforming growth factor-beta stimulates collagen and fibronectin synthesis by human corneal stromal fibroblasts in vitro. *Curr Eye Res* 1993;12:703–709.
65. Hanna KD, Pouliquen Y, Waring GO, et al: Corneal stromal wound healing in rabbits after 193-nm excimer laser surface ablation. *Arch Ophthalmol* 1989;107:895–901.
66. Fantes F, Hanna D, Waring GO, et al: Wound healing after excimer keratomileusis (photorefractive keratectomy) in monkeys. *Arch Ophthalmol* 1990; 108:665–675.
67. Zabel RW, Tuft SJ, Marshall J: Excimer laser photorefractive keratectomy: Endothelial morphology following area ablation of the cornea. *Invest Ophthalmol Vis Sci* 1988;29(Suppl):90.
68. Nishida T, Nakamura M, Murkami J, Mishima H, Otori T: Epidermal growth factor stimulates corneal epithelial cell attachment to fibronectin through a fibronectin receptor system. *Invest Ophthalmol Vis Sci* 1992;33:2464–2469.
69. Chung JH, Fagerholm P: Treatment of rabbit corneal alkali wounds with human epidermal growth factor. *Cornea* 1989;8:122–128.
70. Caporossi A, Manetti C: Epidermal growth factor in topical treatment following epikeratoplasty. *Ophthalmologica* 1992;205:121–124.
71. Scardovi C, De Felice GP, Gazzaniga A: Epidermal growth factor in the topical treatment of traumatic corneal ulcers. *Ophthalmologica* 1993;206:119–124.
72. Cellini M, Baldi A, Caramazza N, DeFelice GP, Gazzaniga A: Epidermal growth factor in the topical treatment of herpetic corneal ulcers. *Ophthalmologica* 1994;208:37–40.
73. Romano A, Peisich A, Wasserman D, Gamus D: Aggravation of herpetic stromal keratitis after murine epidermal grown factor topical application. *Cornea* 1994;13:167–172.
74. Petroutsos G, Sebag J, Courtois Y: Epidermal growth

factor increases tensile strength during wound healing. *Ophthalmic Res* 1986;18:299–300.

75. Sheardown H, Wedge C, Chou L, Apel R, Rootman DS, Cheng YL: Continuous epidermal growth factor delivery in corneal epithelial wound healing. *Invest Ophthalmol Vis Sci* 1993;34:3593–3600.

76. Hommel U, Harvey TS, Driscoll PC, Campbell ID: Human epidermal growth factor. High resolution structure and comparison with human transforming growth factor-α. *J Mol Biol* 1992;227:271–282.

77. Luetteke NC, Qiu TH, Peiffer RL, Oliver PO, Smithies O, Lee DC: TGFa deficiency results in hair follicle and eye abnormalities in targeted and waved-1 mice. *Cell* 1993;73:263–278.

78. Khaw PT, Schultz GS, MacKay SL, et al: Detection of transforming growth factor-alpha messenger RNA and protein in human corneal epithelial cells. *Invest Ophthalmol Vis Sci* 1992;33:3302–3306.

79. Pasquale LR, Dorman-Pease ME, Lutty GA, Quigley HA, Jampel HD: Immunolocalization of TGF-β1, TGF-β2, and TGF-β3 in the anterior segment of the human eye. *Invest Ophthalmol Vis Sci* 1993;34:23–30.

80. Rao RC, Varani J, Soong HK: FGF promotes corneal stromal fibroblast motility. *J Ocul Pharmacol* 1992;8:77–81.

81. Mishima H, Nakamura M, Murakami J, Nishida T, Otori T: Transforming growth factor-beta modulates effects of epidermal growth factor on corneal epithelial cells. *Curr Eye Res* 1992;11:691–696.

82. Phillips GD, Whitehead RA, Stone AM, Ruebel MW, Goodkin ML, Knighton DR: Transforming growth factor beat (TGF-b) stimulation of angiogenesis: An electron microscopic study. *J Submicrosc Cytol Pathol* 1993;25:149–155.

83. Phillips GD, Whitehead RA, Knighton DR: Inhibition by methylprednisolone acetate suggests an indirect mechanism for TGF-b induced angiogenesis. *Growth Factors* 1992;6:77–84.

84. Peters K, Ornitz D, Werner S, Williams L: Unique expression pattern of the FGF receptor 3 gene during mouse organogenesis. *Dev Biol* 1993;155:423–430.

85. de Iongh R, McAvoy JW: Distribution of acidic and basic fibroblast growth factor (FGF) in the foetal rat eye: Implications for lens development. *Growth Factors* 1992;6:159–177.

86. Dabin I, Courtois Y: Acidic fibroblast growth factor overexpression in corneal epithelial wound healing. *Growth Factors* 1991;5:129–139.

87. Vlodavsky I, Fuks Z, Ishai-Michaeli R, et al: Extracellular matrix-resident basic fibroblast growth factor: Implication for the control of angiogenesis. *J Cell Biochem* 1991;45:167–176.

88. Wilson SE, Walker JW, Chwang EL, He YG: Hepatocyte growth factor, keratinocyte growth factor, their receptors, fibroblast growth factor receptor-2 and the cells of the cornea. *Invest Ophthalmol Vis Sci* 1993;34:2544–2561.

89. Adamis AP, Meklir B, Joyce NC: Rapid communication. In situ injury-induced release of basic-fibroblast growth factor from corneal epithelial cells. *Am J Pathol* 1991;139:961–967.

90. Hecpuet C, Morisset S, Lorans G, Plouet J, Adolphe M: Effects of acidic and basic fibroblast growth factors on the proliferation of rabbit corneal cells. *Curr Eye Res* 1990;9:429–433.

91. Rieck P, Assouline M, Savoldelli M, et al: Recombinant human basic fibroblast growth factor (Rg-bFGF) in three different wound models in rabbits: Corneal wound healing effect and pharmacology. *Exp Eye Res* 1992;54:987–998.

92. Mazue G, Bertolero F, Jacob C, Sarmientos P, Roncucci R: Preclinical and clinical studies with recombinant human basic fibroblast growth factor. *Ann NY Acad Sci* 1991;638:329–340.

93. Hoppenreijs V, Pels E, Vrenson G, Felton P, Treffers WF: Platelet-derived growth factor: Receptor expression in cornea and effects on corneal cells. *Invest Ophthalmol Vis Sci* 1993;34:637–649.

94. Talamo JH, Gollamudi S, Green WR, de la Cruz Z, Filatov V, Stark WJ: Modulation of corneal wound healing after excimer laser keratomileusis using topical mitomycin c and steroids. *Arch Ophthalmol* 1991;109:1141–1146.

95. Bergman RH, Spigelman AV: The role of fibroblast inhibitors on corneal healing following photorefractive keratectomy with 193-nm excimer laser in rabbits. *Ophthalmic Surg* 1994;25:170–174.

# section two
Preoperative Evaluation

CHAPTER 4
Corneal Topography in Phototherapeutic Keratectomy

Michael Rogers, Russell McCally, Dimitri T. Azar

HISTORY

It has long been recognized that the first Purkinje image, or corneal light reflex, provides a measure of corneal shape and thus forms a basis for most methods of measuring the corneal surface. The earliest known attempt to measure the curvature of the anterior corneal surface was in 1619 by Father Christopher Scheiner (1573–1650).[1,2] Scheiner compared images reflected from the surface of the cornea with those reflected by glass balls of various diameters. Later, in 1808, David Brewster (1781–1868) used the Purkinje image from a candle flame to observe the shape of a conical cornea.[2,3]

In 1847, Henry Goode described the first keratoscope.[2] It was a small, luminous square held a few inches in front of the eye, allowing him to detect an abnormal corneal shape by observing the corneal reflex. This early keratoscope provided a qualitative measure of corneal deformity. Hermann Ludwig Ferdinand von Helmholtz (1821–1894) is generally credited with the invention of the ophthalmometer, or keratometer, in 1854,[4] although it has also been attributed to Jesse Ramsden (1735–1800).[5] The keratometer made possible the first quantitative measurement of the corneal radius of curvature. Helmholtz's design was improved by Louis Emile Javal (1839–1907) and Hjalmar August Schiøtz (1850–1927) in about 1880.[1]

Working independently of Goode, in 1880 a Portuguese oculist named Antonio Plácido (1840–1916) developed his own hand-held keratoscope for detecting corneal irregularities.[2] *Plácido's disk,* as it came to be called, consisted of a flat disk 23 cm in diameter, with alternating light and dark bands on its surface and a central aperture for viewing the corneal reflex of the rings. In that same year, Plácido developed the first photokeratoscope by using a camera in conjunction with his disk to record the corneal reflex. At about the same time, Javal reported the development of his own keratoscope and photokeratoscope, apparently independently of Plácido.[1,2] Javal went on to present the first known application of photokeratoscopy to the observation of a pathological corneal irregularity.[2] In his review of keratoscopy, Clark attributed the invention of *quantitative* photokeratoscopy to Gullstrand,[6] and he described the major developments from Gullstrand's work up to the end of the 1960s (also noteworthy is Clark's review of keratometry[7]).

In the late 1970s, new techniques were developed to obtain more detailed information about the shape of the cornea. Most of these techniques were enhancements of keratoscopes and photokeratoscopes. Functionally, the principal difference between a keratometer and a keratoscope is in the coverage provided; a keratometer typically uses a simple mire with only two reference points, but a keratoscope usually projects a pattern over a large area of the cornea. Both instruments rely on measuring the reflected image of a target and calculate the radius of curvature based on the size or location of the image.

In 1981, Rowsey and colleagues described a modified photokeratoscope, called the Corneascope, that utilized a comparator to determine the corneal radius of curvature at a point on any of the rings.[8] The comparator allowed the keratoscope photo to be variably enlarged until a point of interest on one of the rings in the photo matched the corresponding point on a standard set of rings. The required magnification was used to determine the radius of curvature at that point. Also in 1981, Doss and colleagues introduced mathematical techniques for analyzing keratoscope photographs to determine the radius of curvature.[9] They used a computer program to calculate the radius from the measured radial distance to each ring in the photo. Their algorithm assumed that the radius of curvature of the cornea at its apex was 7.8 mm, and they calculated the radius of curvature at each ring by constructing a series of arcs tangentially connected at the ring locations. In 1983, Rowsey and Isaac described a technique in which they manually digitized keratoscope photos of calibration spheres and used regression to develop a relationship between the ring locations and the radius of curvature.[10] They incorporated the regression results in a computer program that could be used to analyze digitized keratoscope photos of patients' corneas.

Klyce refined the algorithms developed by Doss and used statistical methods to reduce errors from manual digitization.[11] Klyce's algorithm eliminated the assumption of a 7.8-mm apical radius of curvature by estimating the central radius of curvature from the average radius of the image of the innermost keratoscope ring. His computer program presented results in the form of a three-dimensional representation of the corneal shape. Corneas were also displayed as *distortion plots,* or spherical difference plots, showing how the corneal shape differed from a sphere. Further refinements to the algorithms developed by Doss were published by van Saarloos and Constable.[12] Maguire and Singer worked with Klyce to add a color-coded display showing contours of constant dioptric power.[13] This *isodioptric* display is one of the standard displays in computer-assisted videokeratoscopes now commercially available and is probably the one most widely used. The first integrated, computer-assisted videokeratoscope system was described by Gormley and colleagues in 1988.[14] Typical of the systems now available, it included a lighted target, a charge coupled display video camera to capture the keratoscope image, and computer digitization of the ring locations using an edge detection algorithm.

BASICS OF CORNEAL TOPOGRAPHY

The tear film of the anterior corneal surface acts like a convex mirror to form a virtual, erect image of reflected light. This corneal light reflex is the basis of most topographic measurement methods. For a normal cornea with a central radius of curvature of 7.80 mm, the corneal light reflex is located approximately 3.90 mm posterior to the anterior corneal surface[15] along a normal to the corneal surface.

There are some important facts to remember about the location of the corneal light reflex in relation to the optical landmarks of the cornea. We know that the entrance pupil is the virtual image formed by refracted light from the real pupil.[15] When the eye fixates on a point at optical infinity, the line of sight is defined by the path of a refracted ray passing through the center of the entrance pupil and the center of corneal curvature.[15] The line of sight is generally *not* normal to the cornea. Since corneal topography is increasingly used for planning surgical procedures, topography data should be centered on a useful and easily determined reference point. Uozato and Guyton discussed centering procedures for corneal surgery, and their discussion is applicable here as well.[15] They recommended centering on the entrance pupil while the patient fixates on a target that is made to be coaxial with the observer's line of sight. They also illustrated the centration error that arises from centering on the corneal light reflex instead of the entrance pupil. Topography should be referenced to the center of the entrance pupil, but the concentric rings of a videokeratoscope target center on the corneal light reflex, and the two locations do not, in general, coincide. The offset can be quite significant for an irregular cornea and must be taken into account when using videokeratoscope data.[16] Some videokeratoscope systems have incorporated an algorithm that searches the captured image for the edge of the entrance pupil. This is a step in the right direction, but we have not seen any data on how well the pupil is located.

A distinction should be made between the high point of the cornea and the region of greatest corneal curvature. In an abnormal cornea, the two very likely will not coincide. In the literature, the terms *apex* and *vertex* have been variously used to describe these concepts, leading to some confusion. According to *Webster's Dictionary,* both terms refer to a point on a shape farthest from its base.[17] *Vertex* can also refer to the point where the axis of a curve intersects the curve itself.[16,17] Maloney suggested that the high point of the cornea be denoted the *corneal vertex* and stated that the apex is the region of greatest curvature.[16] However, Waring defined the apex as the high spot of the cornea.[18] We will use *Webster's* definition of apex and will not differentiate between apex and vertex.

Corneal topography is frequently described in terms of radius of curvature. To determine a radius of curvature at a point on a three-dimensional surface, a two-dimensional curve must be defined by intersecting the surface with a plane. This is because the radius of curvature is defined at a point on a two-dimensional plane curve. Such a defining curve and the radius of curvature associated with it are uniquely defined by the orientation of the intersecting plane (except in the special case of a perfect sphere). In describing the optical performance or shape of an aspherical optical element, two perpendicular planes are conventionally defined: the tangential (also called *meridional*) and the sagittal (also called *transverse*) planes.[19,20] A tangential plane is a plane containing both an off-axis object point and the optical axis. The sagittal plane is perpendicular to the tangential plane, and contains the object point and the normal to that object point. There are an infinite number of such planes corresponding to all of the object points in space.

The tangential radius of curvature at a point on a surface is determined as the radius of curvature at that point along a curve defined by the intersection of the tangential plane with the surface; the sagittal radius of curvature is similarly defined. Both radii of curvature lie along the surface normal. For a spherical surface, the tangential and sagittal radii of curvature are identical everywhere, and their centers of curvature are both on the optical axis. However, for an aspherical surface, the tangential and sagittal radii of curvature are independent of one another; they may vary from point to point on the surface; and their centers may both be off of the optical axis. The intersection of the normal to the curve with the optical axis has been referred to as the *axial distance* and has been used to approximate radius of curvature in corneal topography.[21] For a spherical surface, the axial distance is equivalent to the radius of curvature. However, for an aspherical surface, the axial distance may be quite different from any true radius of curvature, whether in the tangential or the sagittal direction.[21]

Radius of curvature, which is an indicator of shape, has been described in terms of dioptric power, an indicator of optical performance, a practice that has led to some confusion.[21] Although paraxial equations, by definition, are valid only near the optical axis, they are frequently used in geometrical optics to estimate optical performance. The paraxial relationship predicts that refractive power is inversely proportional to the radius of curvature. Thus, in the central paraxial region, a surface with a small radius of curvature has more refractive power than one with a larger radius of curvature. However, it is important to note that the paraxial description is valid only in the central 1 to 2 mm of the corneal surface. Outside that region, refractive power is not necessarily inversely proportional to the radius of curvature.[21] The region of interest for corneal topography covers an area of the corneal surface over 8 mm in diameter, well outside the paraxial region, yet corneal topography instruments commonly apply the same Equation 2 over the entire region. To be precise, we use the term *dioptric power* to describe the values displayed by corneal topography "power" maps based on corneal shape, and our use of the term *refractive power* refers strictly to optical performance. Dioptric power maps should not be interpreted, even qualitatively, as refractive power maps.

VIDEOKERATOSCOPY

Although several other methods for measuring corneal topography have been reported, some of which are now commercially available, Plácido's disk-based topography systems still dominate in clinical practice.

As we have already pointed out, the paraxial description of refractive power is an approximation valid only near the optical axis. However, videokeratoscopes are used to measure a region that is approximately 8 mm in diameter. Roberts showed that videokeratoscopes do not display refractive power accurately over this entire region and may produce incorrect qualitative patterns.[21] Thus, dioptric power data produced by videokeratoscopes should be considered a description of shape only, *not* of refractive power. Both corneal shape and refractive power are meaningful quantities for understanding optical performance, but the distinction is essential for correct interpretation of corneal measurements.

Computer-assisted videokeratoscope systems that are commercially available today use different proprietary algorithms to compute the radius of curvature. All suffer from the fact that there is insufficient information in the two-dimensional image of the corneal reflex to uniquely determine the three-dimensional shape of the cornea.[22] Many videokeratoscopes overcome the ambiguity of surface reconstruction by calibrating the mire images on a series of standard spherical surfaces with known radii of curvature.[23,24] An internal table of the mire locations for the calibration surfaces is created. When a measurement is made of a cornea, the measured locations of the mire images are compared to those obtained during calibration to determine the radius of curvature at each point. This technique is essentially an adaptation of the method proposed by Rowsey et al.[8] Measurements of aspherical surfaces made with this method are subject to error due to the spherical bias of the calibration procedure.[22,23]

There are several consequences of spherical bias that merit further discussion. When a light ray is reflected by a surface, it lies in the same plane as the incident ray and the surface normal. All the surface normals of a spherical optical surface pass through the optical axis, and the center of curvature at each surface point is also located on the optical axis. For such a surface, a reflected ray and the center of curvature are contained in the same plane. However, a spherical surface is a special case; for the more general case of an aspherical surface, the center of curvature and the radius of curvature must be defined in terms of planes intersecting the surface at the point of interest, and they will vary with the orientation of such planes (conventionally, the tangential and sagittal planes). In any case, for an aspherical surface, the plane defined by the reflected ray and the surface normal may not contain the optical axis. A spherically biased videokeratoscope cannot resolve this type of optical performance, so some information about an irregular surface is lost. To completely characterize the surface normal, which ultimately defines optical performance, both the sagittal and the tangential radii of curvature are necessary.

Systems that are spherically biased do not determine the tangential or instantaneous radius of curvature because they assume that the center of curvature lies on the optical axis.[23] Roberts quantified the error in the dioptric power determinations of the EyeSys Corneal Analysis System (EyeSys Laboratories, Houston, TX) by comparing measured and theoretically calculated curvature for a sphere and an ellipsoid with known dimensions.[12] In that study, the EyeSys videokeratoscope was characterized as spherically biased, since it calculates the radius of curvature by comparing the keratoscope image's dimensions with those in a look-up table. The videokeratoscope measured the sphere under conditions of perfect alignment to within 0.10 D of the "true" paraxial power. Comparison of measured and theoretical paraxial dioptric powers for the ellipse showed that at the inner four rings the measurement error was less than 0.25 D, but it increased peripherally to a maximum of 3.22 D at the outermost ring. Comparisons between aligned and misaligned measurements indicated that the misalignment error was small compared to the inherent error of the algorithm itself. However, comparison of the measured results to calculations using a formula based on axial distance showed that the instrument accurately reproduced the axial distance rather than the radius of curvature. Similar results were obtained for the Topographic Modeling system (Computed Anatomy, New York, NY).[25]

Currently available Plácido disk-based topography systems also are incapable of measuring the sagittal radius of curvature because, with a target consisting of a series of concentric rings, it is not possible to measure distortion of the virtual image in the sagittal direction.[23] A rectilinear keratoscope target such as that proposed by Shimmick and Munnerlyn may be able to overcome this shortcoming, since it would be able to resolve image displacement in two orthogonal directions.[26] In a new product under development by EyeSys Laboratories, a pattern of spokes is superimposed on the Plácido disk target to produce a "checkerboard" Plácido. The sagittal radius of curvature could be measured by analyzing the distortion of the reflected checkerboard pattern in the sagittal direction.

An algorithm for computing corneal curvature that avoids the assumption of corneal sphericity and that allows that instantaneous center of curvature to lie off the optical axis (but still in the tangential plane) was introduced by Wang, Rice, and Klyce in 1989.[22] They found that the maximum error for measurement of an ellipsoid was reduced from 8% with a spherically biased algorithm to 2% with the new method. Increased computation time and sensitivity to noise have prevented its implementation so far. Several manufacturers of corneal topography devices have developed, or are in the process of developing, algorithms based on tangential or instantaneous radius of curvature rather than axial distance. Systems in which the tangential or instantaneous radius of curvature is currently available as an output include the EyeSys Corneal Analysis System, the Keratron Corneal Imaging Analyzer (Optikon 2000, Milan, Italy), and the Alcon EyeMap EH290 Corneal Topographer (Irvine, CA). A similar algorithm is currently under development for the Topographic Modeling System. However, no data have yet been published on the accuracy of any of these algorithms in measuring radius of curvature. These algorithms will potentially determine the tangential radius of curvature more accurately than a spherically biased algorithm, but they still neglect the sagittal radius of curvature.

CLINICAL APPLICATIONS

Corneal topography is helpful in a number of clinical situations, including evaluation of patients with astigmatism. Figure 4.1 illustrates with-the-rule astigmatism in a patient who underwent cataract surgery. The superior corneal incision was sutured with loose 10-0 Nylon sutures to minimize postoperative astigmatism. The corneal topography map illustrates the relative flattening of the superior hemi-meridian compared to the inferior hemi-meridian with persistence of with-the-rule astigmatism. Figure 4.2 illustrates against-the-rule astigmatism with a steep horizontal meridian and the typical bow-tie appearance of the corneal topography. The two patients illustrated in Figures 4.1 and 4.2 have regular astigmatism; those shown in Figure 4.3 and 4.4 have keratoconus.

The use of corneal topography in phototherapeutic keratectomy (PTK) is especially valuable in detecting treatment decentration (Figure 4.5). The hyperopic shift associated with PTK is illustrated in Figures 4.6 and 4.7. The presence of a central island is not unusual after PTK, as shown in Figure 4.7A and 4.7B. Wound healing following PTK can result in significant alterations of the corneal surface which can be detected by serial topographical maps of the cornea after PTK (Figure 4.8).

Although PTK may induce surface irregularities, penetrating keratoplasty may result in high degrees of astigmatism (Figure 4.9). Corneal topography is very helpful in planning surgical correction of the induced astigmatism after corneal surgery. The most common indication of corneal topography is perhaps in the preoperative and postoperative evaluation of refractive surgical patients. Myopes show significant flattening of the central cornea following incisional, laser, and lamellar refractive procedures, as illustrated in Figure 4.10. The corneal topographical changes after PTK are illustrated in

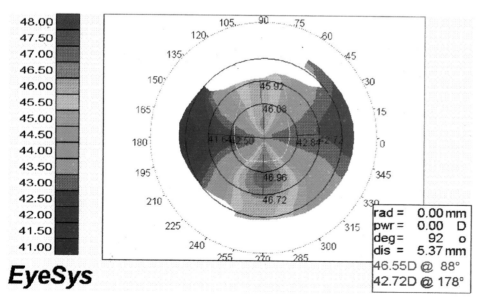

Figure 4.1. With-the-rule astigmatism.

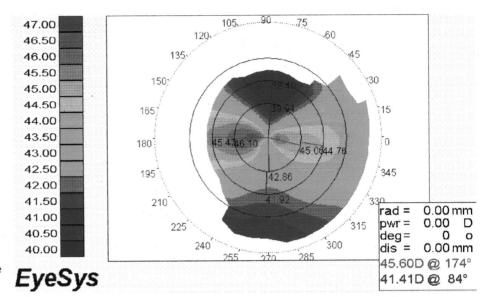

Figure 4.2. Against-the-rule astigmatism.

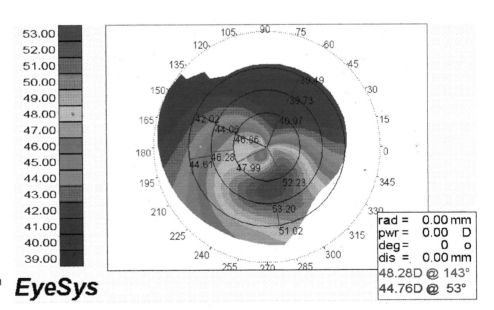

Figure 4.3. Keratoconus with typical inferior steepening.

Corneal Topography in Photothermapeutic Keratectomy

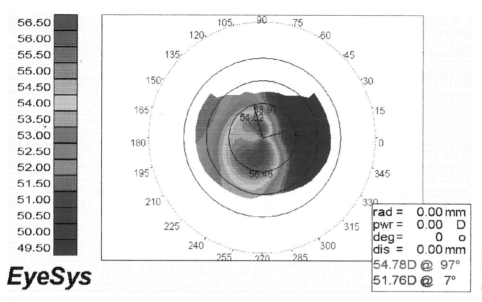

Figure 4.4. Keratoconus with inferiotemporal and superiotemporal steepening.

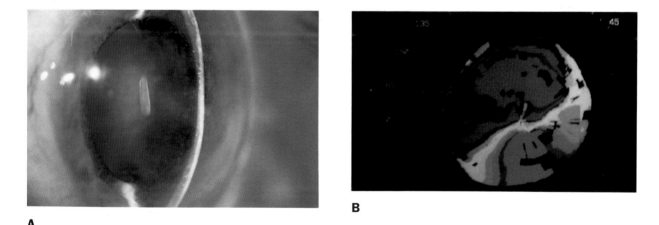

Figure 4.5. A slit-lamp photograph (A) and the corneal topography (B) of a 45-year-old man 12 months after PTK for lattice dystrophy. Note the area of superior fattening in the treatment zone.

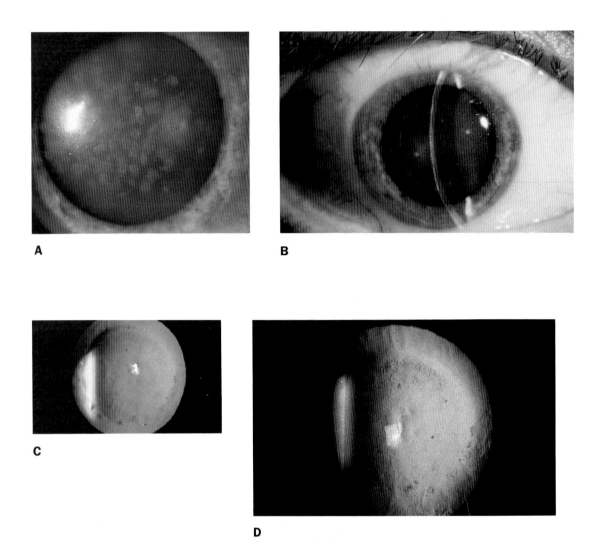

Figure 4.6. A 73-year-old woman with granular dystrophy who underwent PTK. Right eye: *A:* Preoperative appearance of the cornea with visual acuity of 20/600. *B, C,* Appearance 2 months after PTK by direct illumination and retroillumination, respectively, with visual acuity of 20/400. *D:* Appearance 18 months after PTK with visual acuity of 20/200.

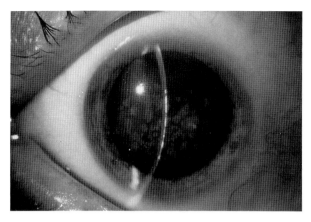
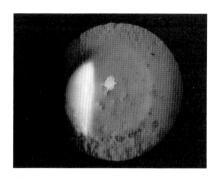
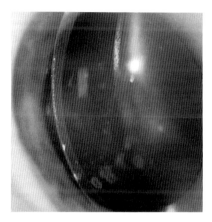
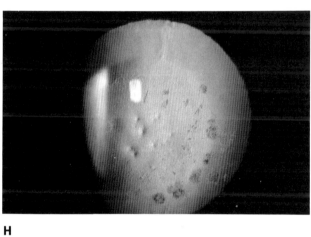

Figure 4.6. Left eye: *E:* Preoperative appearance with visual acuity of 20/59. *F:* Appearance 9 months after PTK. Visual acuity was 20/30. *G, H:* Appearance 18 months after PTK by direct illumination and retroillumination, respectively, with similar visual acuity.

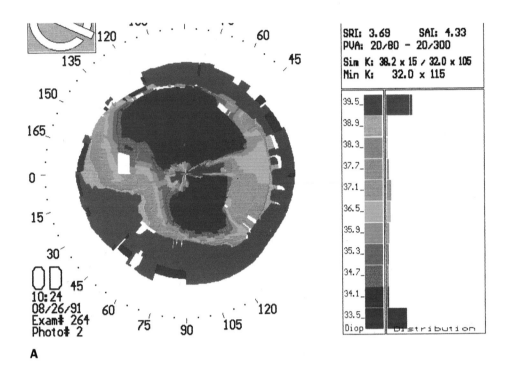

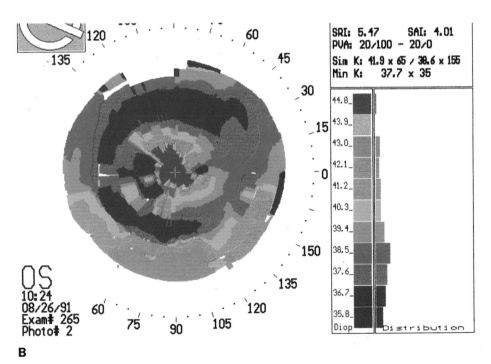

Figure 4.7. *A:* Corresponding postoperative topography of the right eye of the patient shown in Figure 4.6 15 months after PTK showing a corneal flattening resulting in hyperopic shift. *B:* Postoperative topography of the left eye.

Corneal Topography in Phototherapeutic Keratectomy

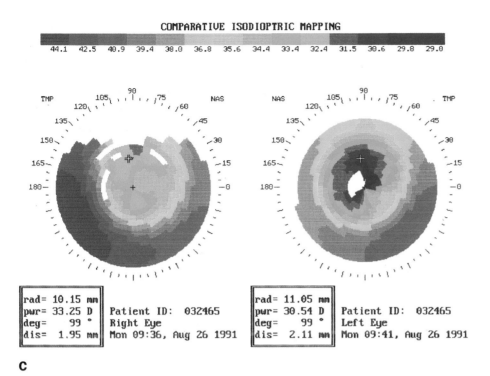

Figure 4.7. *C:* Comparative isodioptric mapping the two eyes.

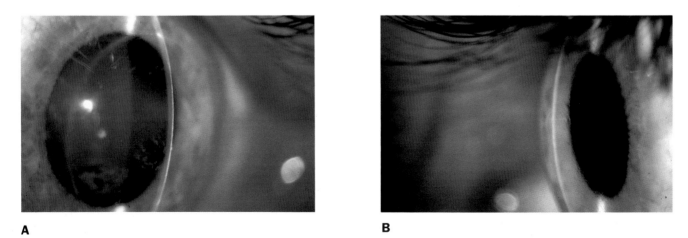

Figure 4.8. A 31-year-old man with recurrent lattice dystrophy who underwent PTK and later photoastigmatic keratectomy (PAK) for correction of myopia and astigmatism in the right eye. The patient is one of the brothers shown in Figure 4.6. *A:* A slit lamp photograph of the eye prior to PTK. *B:* Appearance 11 months after PAK.

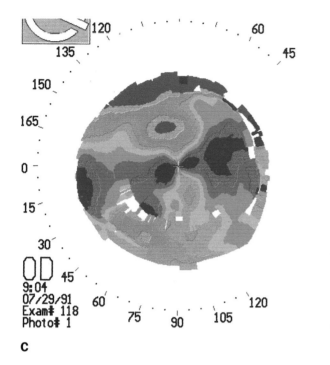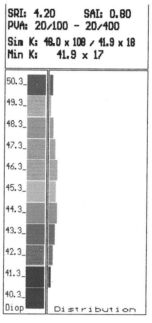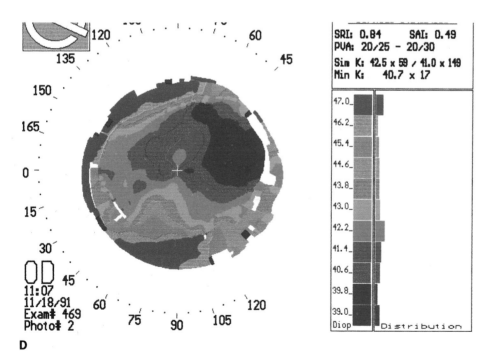

Figure 4.8. *C, D:* Corresponding corneal topography 1 and 11 months postoperatively showing significant changes in corneal surface during wound healing following PTK.

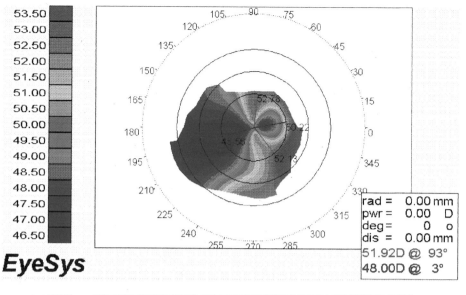

Figure 4.9. Irregular astigmatism following penetrating keratoplasty at the 12 o'clock and 5 o'clock areas represents steepening that corresponds to tight sutures in contrast to the 9 o'clock flat region.

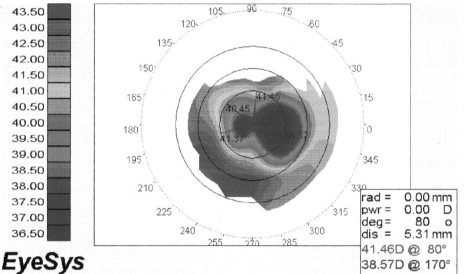

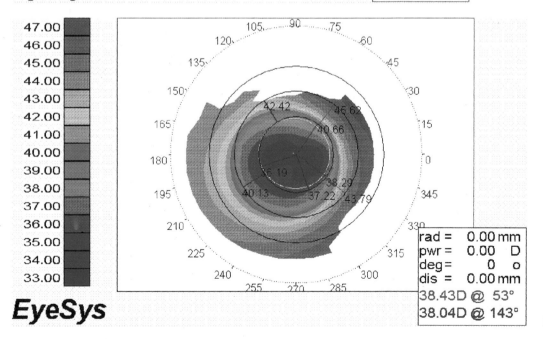

Figure 4.10 Central corneal flattening following radial keratotomy. (A) and keritomileusis surgery (B).

chapter 10; and the corneal alterations after PRK are discussed in chapter 11. Chapter 12 describes the pathogenesis of central islands after PRK and describes the patterns and management of topographical changes after PRK including those of decentered PRK treatments.

The use of PTK to correct corneal irregularities may be enhanced in the future by using the height data obtained from corneal topographical maps to guide the laser and smooth the corneal surface. Current limitations in achieving this goal include (1) unreliability of sag values obtained from spherically based topographical data; (2) difficulty of treatment centration; and (3) limitations of tracking systems. With future research in this area, it is possible to overcome these limitations.

REFERENCES

1. Gorin G: *History of Ophthalmology*. Wilmington, DE, Publish or Perish, 1982.
2. Levene JR: The true inventors of the keratoscope and photo-keratoscope. *Br J Hist Sci* 1965;2:234–342.
3. Levene JR: Sir David Brewster (1781–1868) and the clinical detection of corneal abnormalities. In *12th International Congress on the History of Sciences*, Vol 8. Paris, 1968, pp 105–109.
4. Southall JPC: *Introduction to Physiological Optics*. New York, Dover, 1937.
5. Mandell RB: Jesse Ramsden: Inventor of the ophthalmometer. *Am J Optom* 1960;37:633.
6. Clark BAJ: Conventional keratoscopy—a critical review. *Aust J Optom* 1973;56:145–155.
7. Clark BAJ: Keratometry: A review. *Aust J Opthom* 1973;56:94–100.
8. Rowsey JJ, Reynolds AE, Brown R: Corneal topography. *Arch Ophthalmol* 1981;99:1093–1100.
9. Doss JD, Hutson RL, Rowsey JJ, Brown DR: Method for calculation of corneal profile and power distribution. *Arch Ophthalmol* 1981;99:1261–1265.
10. Rowsey JJ, Isaac MS: Corneoscopy in keratorefractive surgery. *Cornea* 1983;2:133–142.
11. Klyce SD: Computer assisted corneal topography: High resolution graphic presentation and analysis of keratoscopy. *Invest Ophthalmol Vis Sci* 1984;25:1426–1435.
12. van Saarloos PP, Constable IJ: Improved method for calculation of corneal topography for any photokeratoscope geometry. *Optom Vis Sci* 1991;68:960–965.
13. Maguire LJ, Singer DE, Klyce SD: Graphic presentation of computer-analyzed keratoscope photographs. *Arch Ophthalmol* 1987;105:223–230.
14. Gormley DJ, Gertsen M, Koplin RS, Lubkin V: Corneal modeling. *Cornea* 1988;7:30–35.
15. Uozato H, Guyton DL: Centering corneal surgical procedures. *Am J Ophthalmol* 1987;103:264–275.
16. Maloney RK: Corneal topography and optical zone location in photorefractive keratectomy. *Refract Corneal Surg* 1990;6:363–371.
17. Mish FC (ed): *Webster's Ninth New Collegiate Dictionary*. Springfield, MA: Merrian-Webster, 1984.
18. Waring GO: Making sense of keratospeak II: Proposed conventional terminology for corneal topography. *Refract Corneal Surg* 1989;5:362–367.
19. Bennett AG: Aspheric contact lens surfaces. *Ophthalmol Opt* 1968;8:1037–1040, 1297–1300, 1311; and 9:222–230.
20. Jenkins FA, White HE: *Fundamentals of Optics*, 3rd ed. New York, McGraw-Hill, 1957.
21. Roberts C: The accuracy of 'power' maps to display curvature data in corneal topography systems. *Invest Ophthalmol Vis Sci* 1994;35:3525–3532.
22. Wang J, Rice DA, Klyce SD: A new reconstruction algorithm for improvement of corneal topographical analysis. *Refract Corneal Surg* 1989;5:379–387.
23. Roberts C: Characterization of the inherent error in a spherically-biased corneal topography system in mapping a radially aspheric surface. *J Refract Corneal Surg* 1994;10:103–116.
24. McCarey BE, Zurawski CA, O'Shea DS: Practical aspects of a corneal topography system. *CLAO J* 1992;18:248–254.
25. Roberts C: Analysis of the inherent error of the TMS-1 Topographic Modeling System in mapping a radially aspheric surface. *Cornea* in press.
26. Shimmick J, Munnerlyn C: Corneal analysis with a rectilinear photokeratoscope. In *Ophthalmic and Visual Optics Technical Digest*. Washington, DC: Optical Society of America, 1992, pp 2–3.

CHAPTER 5

Preoperative and Postoperative Protocols

Christopher J. Rapuano

Attempting to set standard protocols regarding any preoperative, treatment, or postoperative aspects of excimer laser phototherapeutic keratectomy (PTK) is difficult given the wide variety of pathology treated with this technique. Given that caveat, this chapter will combine the essentials from numerous protocols with several years of clinical practice in performing PTK to present a coherent guide for preoperative and postoperative care.[1–4]

PREOPERATIVE EVALUATION

A thorough systemic and ocular history is important. Conditions which affect healing, especially collagen vascular disorders, and evidence of active inflammation are often contraindications to PTK. A complete ophthalmic evaluation, including uncorrected visual acuity, manifest refraction, intraocular pressure measurement, keratometry readings and/or corneal topography (if possible), and a dilated fundus examination is essential for any patient considering excimer laser surgery. To be a candidate for excimer laser PTK, the patients' symptoms need to be consistent with the results of the corneal examination. For example, a patient with symptoms of recurrent erosion should have localizable surface pathology explaining the painful episodes. Also, patients with decreased vision should have corneal pathology consistent with the amount of visual impairment. In general, the best candidates for excimer laser PTK are patients with pathology in the anterior 5–20% of the cornea.

The best way to determine the depth of corneal pathology in most patients is by a thorough slit lamp examination. The percentage of the cornea involved can be closely approximated. In addition, an absolute depth—for example, 50 μm versus 75 μm—can be estimated at the slit lamp. Ultrasonic pachymetry is then used to determine the thickness of the entire cornea in the area of the pathology. The actual depth of the corneal opacity can then be roughly calculated. If the cornea is 550 μm thick and the pathology reaches 20% thickness, then the opacity is 110 μm deep. In practice, the depth of pathology is often somewhat difficult to assess accurately at the slit lamp, even by experienced clinicians. Some investigators have found the optical pachymeter helpful in determining the depth of corneal pathology more accurately.[5] Optical pachymetry, however, is not widely available. Ultrasonic biomicroscopy, a relatively new technique which uses high-frequency ultrasound to image the anterior segment, may prove valuable in measuring depth of corneal opacities in the future.[5–9]

Another useful tool in the preoperative evaluation of PTK patients is computerized corneal topography. Placido disc techniques (e.g., EyeSys Technologies, Houston, TX; and Computed Anatomy, New York, NY) can supply a great deal of information regarding regular and irregular astigmatism. However, a relatively smooth surface is necessary to obtain useful data using these machines. Other, non-Placido disc techniques (e.g., PAR Vision Systems Corp., New Hartford, NY) provide physical elevation data along with refractive information. Often the refractive and elevation data are helpful in planning the exact surgical technique. In addition, comparison of preoperative and postoperative corneal curvatures and elevations is indispensable in evaluating, modifying, and perfecting PTK techniques.

Table 5.1. Preoperative and Postoperative Examinations

Complete ocular and systemic history
Visual acuity without correction
Manifest refraction
Pupillary evaluation
Extraocular motility
Confrontational visual fields
Intraocular pressure
Slit lamp evaluation
   Size and depth of corneal pathology
   Schirmer's test, as needed
   Corneal sensation, as needed
Keratometry
Pachymetry (ultrasonic and/or optical)
Computerized corneal topography
Slit lamp photography, if available
Ultrasound biomicroscopy, as needed
Dilated fundus evaluation

SURGICAL PREPARATION

Laser Setup

As in all surgery using the excimer laser, the machine needs to be finely tuned and calibrated prior to use. Current excimer laser units require very high maintenance. They need a constant supply of gases, which have to be checked a few days before surgery to make sure that there is a sufficient amount for all the procedures planned. On the day of surgery, the laser beam's power and quality need to be evaluated to make certain that they are the same as predicted. In this sense, the excimer laser is different from most other lasers used in ophthalmology. When a surgeon performs argon laser trabeculoplasty, he or she can adjust the power of the beam after the first shot or two, depending on the response of the tissue. The same is true of yttrium-aluminum-garnet (YAG) laser posterior capsulotomy and argon laser retinal photocoagulation. In contrast, the excimer laser beam is not adjusted intraoperatively and consequently needs to be perfectly calibrated preoperataively.

Laser calibration is different for each laser unit. The fluence needs to be adjusted to the specifications. The surgeon then ablates a material which is used to calibrate the laser power output. Typically, an output is set, the laser fired, and the result measured. When the result does not equal the output setting, the machine is adjusted until the predicted and achieved results are identical. In addition, the quality of the excimer laser beam is evaluated. Material such as a polymethylmethacrylate block or special film is ablated and checked to make certain that the laser beam's shape and uniformity are perfect.

Preoperative adjustment of the centration of the excimer laser beam is extremely important, as it is quite difficult to determine the exact location of the beam (to within 0.25–0.5 mm) during corneal ablation. While numerous methods exist, a simple way to determine the laser beam's location is to decrease the spot size to 2 mm and ablate a piece of colored magazine paper. Within a few pulses, the print is ablated and the ablation area is checked against the centration reticle in the operating microscope oculars. The x and y axes of the beam location are then adjusted until the beam is perfectly centered. This step is so important that it should be performed prior to each surgical procedure.

Surgical Planning

Immediately before surgery, the patient is examined at the slit lamp to confirm the exact location of the pathology and plan the surgical technique. In general, smooth stromal opacities (Figure 5.1) (e.g., corneal dystrophies) are treated differently from elevated nodules (Figure 5.2) (e.g., Salzmann's degeneration) and other surface irregularities (Figure 5.3) (e.g., certain corneal scars) (Figure 5.4).

Once the patient has been examined, a topical anesthetic (e.g., proparacaine or tetracaine) is placed in both the operative and fellow eyes. Some surgeons find that a preoperative topical nonsteroidal anti-inflammatory drop (e.g., diclofenac or ketorolac) decreases postoperative pain. The fellow eye is then patched. The patient is placed on the operating chair and positioned under the laser. A head rest (e.g., Vac-Pac, Olympic Medical, Seattle, WA) is helpful in stabilizing the patient's head. Strict care needs to be taken to ensure that the patient's head is level and the operative eye is centered under the operating microscope. A sterile eyelid speculum is placed. The patient is asked to look at the fixation light in the microscope. Magnifi-

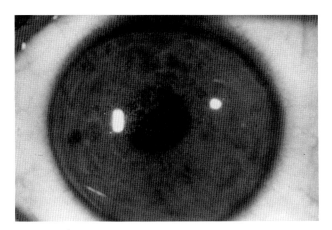

Figure 5.1. Smooth stromal opacity in a patient with superficial variant of granular dystrophy. In general, a large (e.g., 6-mm) circular ablation is performed without mechanical epithelial removal centered over the visual axis.

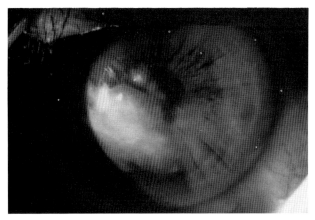

Figure 5.3. A dense, irregular corneal scar remains from a chemical injury many years earlier. Masking agents and selective ablation should be used to smooth the surface and remove the opacity. Care needs to be taken not to ablate too deeply or excessive flattening and hyperopia will be induced.

cation needs to be adjusted according to the directions of the laser manufacturer. PTK should be avoided in patients with surface irregularities resulting from endothelial decompensation (Figure 5.5).

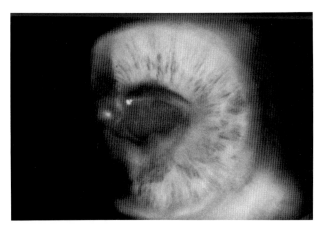

Figure 5.2. Dense elevated corneal opacity in a patient with Salzmann's nodular degeneration. The epithelium over the elevated opacity is usually removed mechanically. A masking agent is used to fill depressions in order to smooth the surface more effectively. The treatment is not necessarily centered over the entrance pupil but rather on the pathology.

In patients with smooth epithelial surfaces with anterior corneal opacities, such as granular dystrophy or Reis-Bücklers' dystrophy, the goal is to reproduce the smooth surface of the cornea at a deeper level, below the bulk of the opacity. In these cases, mechanical epithelial debridement is usually not performed. Instead, the laser beam, typically a circle with a large diameter such as 6.0 or 6.5 mm, is used to ablate the epithelium and stroma. Ablations of this type are centered on the entrance pupil, even if the pathology is not centered exactly over the pupil. The ablation depth is set to approximately 75% of the predicted depth of the pathology. Once this first ablation is completed, the patient is examined at the slit lamp to determine whether, and if so how much, additional ablation needs to be performed. This "ablate and check" procedure is continued until the desired amount of cornea is removed.

In patients with elevated opacities, such as Salzmann's degeneration or a keratoconus nodule, the object is to ablate the elevated area without removing relatively normal surrounding and underlying tissue. One way in which this goal is achieved is by removing the epithelium from the elevated nodule while leaving the surrounding epithelium in place. The remaining epithelium acts as a masking barrier to the laser beam, preventing inadvertent removal of surrounding tissue. Liquid masking

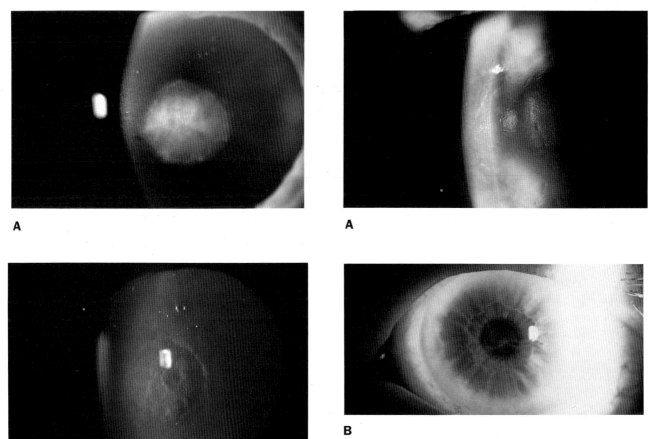

Figure 5.4. Appearance of an eye with a corneal scar after corneal ulceration (A) by direct illumination and (B) by retroillumination.

Figure 5.5. A patient with recurrent erosion syndrome secondary to Fuch's dystrophy. (A) A slit lamp photograph of the right eye showing epithelial irregularities. (B) A photograph of the contralateral eye showing subepithelial scarring and deep folds of epithelium. PTK is contraindicated in such patients.

agents, such as Tears Naturale II (Alcon, Fort Worth, TX) or various concentrations of methylcellulose [½% carboxymethylcellulose (Refresh plus, Allergan, Irvine CA) or 1% carboxymethylcellulose (Celluvisc, Allergan, Irvine CA).], can also be used to prevent ablation where not desired. The masking agent is placed in the "valleys" so that the "mountain tops" can be ablated by the laser. When the epithelium and various solutions are used as masking agents, the valleys can be protected so that the elevated areas can be ablated down to achieve a smoother surface. The treatment of irregular and/or elevated corneal opacities is often not centered on the visual axis but rather on the elevated pathology. In general, small spot-sized ablations are used at first to reduce the elevations, and larger spot sizes are used later in the treatment to smooth the surface. Sharp transition zones should be avoided. Many surgeons find that gentle rocking of the patient's head aids in creating a smoother transition zone between treated and untreated areas of the cornea. Here too, the patient is examined at the slit lamp during the procedure to evaluate the effect of the laser and plan additional treatment. As discussed elsewhere, the deeper the ablation, the greater the tendency to induce corneal flattening and hyperopia.

POSTOPERATIVE REGIMEN

The immediate postoperative regimen depends on the particular treatment performed. If a discrete nodule was removed, leaving a small residual epithelial defect, only antibiotic ointment (e.g., erythromycin qid) may be used. Pressure patching is an option if the surgeon believes that the patient would be more comfortable. If a large epithelial defect remains after PTK, a cycloplegic agent (e.g., scopolamine 1/4%) is often utilized. An antibiotic drop (e.g, ciprofloxacin, ofloxacin, or tobramycin) and a nonsteroidal anti-inflammatory agent (e.g., ketorolac or diclofenac) are typically given. Then a bandage soft contact lens (e.g., Seequence 2, or soflens 66 (Bausch & Lomb, Inc., Rochester, NY or Protek (Ciba) is placed. The patient is reexamined at the slit lamp to make sure that the bandage soft contact lens centers and moves well. If it does not, a contact lens with different parameters may be used. Alternatively, antibiotic ointment (e.g., erythromycin) or antibiotic-steroid combination (e.g. Tobradex) ointment and a pressure patch are placed. Patients with a bandage soft contact lens are instructed to use the antibiotic drop four times a day. Oral pain medication (e.g., acetaminophen with codeine, hydroxycodone, or demerol with promethazine) should be prescribed as needed. Patients often find ice compresses over the eye effective in decreasing discomfort.

Patients should be seen every day or two until the epithelial defect heals. The bandage soft contact lens is removed and replaced with a new lens only if it is tight, damaged, or soiled. The epithelial defect typically heals within 2–6 days, depending on the size of the original defect.

The edges of the defect can usually be seen through the contact lens. One should avoid moving the lens whenever possible, so that epithelial resurfacing is not disturbed. Fluorescein drops may be used to visualize the defect, but are often unnecessary. Staining of the lens may result from the use of fluorescein, but does not seem to interfer with the lense or with re-epithelialization.

Once the epithelial defect has healed, the antibiotic drops are discontinued. Some surgeons then treat with a mild topical steroid (e.g., fluorometholone 0.1% qid), tapering over 2–4 months, while others do not. Generally, an antibiotic ointment (e.g., erythromycin) is used at bedtime for several months.

POSTOPERATIVE EXAMINATION

After the initial epithelial healing phase, patients are usually seen 1 month and 2 to 3 months postoperatively. At these postoperative visits, information similar to that obtained at the preoperative examination is gathered. Specifically, visual acuity without correction, manifest refraction, intraocular pressure (especially if topical steroids are used),

Table 5.2. Postoperative Regimen

Immediately after surgery
  Cycloplegic (e.g., scopolamine 1/4%).
  Nonsteroidal anti-inflammatory agent (e.g., diclofenac or ketorolac).
  Antibiotic drop (e.g., gentamicin ofloxacin or ciprofloxacin) and a bandage soft contact lens (e.g., Seequence 2, Protek [Ciba], or Soflens 66 [Bausch and Lomb])
  or
  Antibiotic ointment (e.g., erythromycin), with or without a pressure patch.

First postoperative days (until reepithelialization)
  Nonsteroidal anti-inflammatory agent (e.g., diclofenac or ketorolac) two to four times a day
  Antibiotic drop (e.g., gentamicin) four times a day
  Remove the bandage soft contact lens on postoperative day 2 or 3. If a large epithelial defect remains, then a new bandage contact lens can be placed and removed 2–3 days later.

After reepithelialization occurs
  Antibiotic ointment (e.g., erythromycin) at night
  Mild steroid drop (e.g., fluorometholone) four times a day, slowly tapered over 1–5 months
  Artificial tears as need for comfort

keratometry readings, and computerized corneal topography should be obtained. This information is both helpful in evaluating the patient's current status and invaluable in assessing the cornea's response to the surgery. The success of a specific smoothing agent or method, or the amount of induced hyperopia, are necessary data to improve a surgeon's PTK technique.

The extent of future postoperative examinations is determined by how the patient is doing and the surgeon's interest in additional follow-up information. As surgeons become more experienced with PTK, they can modify the preoperative and postoperative protocols to satisfy their needs.

REFERENCES

1. Stark WJ, Chamon W, Kamp MT, Enger CL, Rencs EV, Gottsch JD: Clinical follow-up of 193-nm ArF excimer laser photokeratectomy. *Ophthalmology* 1992;99:805–812.
2. Rapuano CJ, Laibson PR: Excimer laser phototherapeutic keratectomy. *CLAO J* 1993;19:235–240.
3. Rapuano CJ, Laibson PR: Excimer laser phototherapeutic keratectomy for anterior corneal pathology. *CLAO J* 1994;20:253–257.
4. Orndahl M, Fagerholm P, Fitzsimmons T, Tengroth B: Treatment of corneal dystrophies with excimer laser. *Acta Ophthalmol* 1994;72:235–240.
5. Stark WJ, Gilbert ML, Gottsch JD, Munnerlyn C: Optical pachometry in the measurement of anterior corneal disease: An alternative tool for phototherapeutic keratectomy. *Arch Ophthalmol* 1990;108:12–13.
6. Reinstein DZ, Silverman RH, Trokel SL, Allemann N, Coleman DJ: High-frequency ultrasound digital signal processing for biometry of the cornea in planning phototherapeutic keratectomy (letter). *Arch Ophthalmol* 1993;111:430–431.
7. Allemann N, Chamon W, Silverman RH, et al: High-frequency ultrasonic quantitative analyses of corneal scarring following excimer laser keratectomy. *Arch Ophthalmol* 1993;111:968–973.
8. Pavlin CJ, Harasiewicz K, Foster FS: Ultrasound biomicroscopic assessment of the cornea following excimer laser photokeratectomy. *J Cataract Refract Surg* 1994;20:206–211.
9. Silverman RH, Rondeau MJ, Lizzi FL, Coleman DJ: Three-dimensional high frequency ultrasonic parameter imaging of anterior segment pathology. *Ophthalmology* 1995;102:837–843.

section three
PTK for Corneal Disorders

CHAPTER **6**

Anterior Corneal Dystrophies: Clinical Features

Suhas Tuli, Shu-Wen Chang, Walter Stark, Dimitri T. Azar

INTRODUCTION

Phototherapeutic keratectomy (PTK) can be the procedure of choice for several corneal dystrophies. Not all patients with corneal dystrophies may benefit from PTK. The purpose of this chapter is to review anterior corneal dystrophies of importance for PTK. Corneal dystrophies are primary, hereditary disorders that occur bilaterally, affect the central cornea, and are not associated with inflammation. They are usually autosomal dominant, with variable expressivity. Recent attempts to classify corneal dystrophies on a genetic basis seem promising. However, the current standard classification scheme, based on the affected layers of the cornea, is useful in grouping corneal dystrophies into distinct phenotypic categories and in predicting the response to PTK, even if these categories are genetically related. Patients with superficial stromal and epithelial basement membrane dystrophies may respond well to PTK. Patients with recurrent granular or lattice dystrophy in a graft have relatively superficial lesions and are especially suitable for treatment with PTK.

EPITHELIAL DYSTROPHIES

Juvenile Hereditary Epithelial Dystrophy (Meesmann's Dystrophy)

First described clinically by Pameijer in 1935 and histopathologically by Meesmann and Wilke in 1939, this rare dystrophy is autosomal dominant, with incomplete penetrance and variable expressivity.[1–5] A recessive form has also been reported.

Meesmann's dystrophy can be observed early in life as a bilateral, symmetric condition involving the interpalpebral zone. Slit lamp examination reveals multiple small, refactile cysts in the epithelium, especially on retroillumination. Positive staining with fluorescein may occur. Near the horizontal limbus, cysts form in whorls or wedge-shaped clusters. Refractile lines may be formed by coalescence of cysts. Patients experience photophobia and pain because of recurrent erosions later in life. The most important symptom requiring surgical intervention is blurred vision. Visual acuity, however, remains good in most cases. Visual disturbances occur due to irregularity in the corneal surface and mild opacification.

The epithelial cysts seen in this condition may be confused with the epitheliopathy that develops in contact lens wearers, toxicity from local anesthetics, vernal conjunctivitis, meibomitis, or dry eyes. These conditions, however, are nonfamilial. Map-dot fingerprint dystrophy is another differential diagnosis, but the differentiation lies in the uniform, diffuse distribution of Meesmann's dystrophy.

Pathology

The basal epithelium shows accumulation of an abnormal intracytoplasmic "peculiar" substance, which stains with periodic acid–Schiff (PAS) reagent and with Hale's colloidal iron.[6] The cysts correspond to areas of cell degeneration surrounded by adjacent cells. There is thickening of the basement membrane[1,2] and an increase in the glycogen content of the basal epithelial cells.[6,7] The increased glycogen content is believed to represent increased cell turnover.[8]

Ultrastructurally, the epithelium shows degenerative changes including intense vacuolation.[9] Cysts have a corrugated wall consistent with acantholysis. An electron-dense, fibrillogranular, "peculiar" substance is found in the cytoplasm of affected cells.

Some believe that the abnormal material is derived from tonofilaments and is closely related to the desmosomes. Others have found electron-dense bodies similar to lysosomes in the epithelial cells. The basement membrane is thickened and appears to have two zones. The thicker one is rich in collagen fibrils mimicking abnormal anchoring fibrils. Frequent intercalated fibroblasts are seen in the thinner zone, which has few fibrils and may represent a repair phenomenon.

Treatment

Treatment of Meesmann's dystrophy is rarely required. It includes symptomatic management of recurrent erosions with hypertonic saline, therapeutic contact lenses, or epithelial debridement. Debridement of the epithelium is usually followed by reappearance of symptoms. Superficial keratectomy is curative but is rarely warranted. In recent years, phototherapeutic keratectomy (PTK) with the excimer laser has emerged as a viable alternative for the treatment of recalcitrant, recurrent erosions. With PTK the epithelial defect sites are treated to a depth of 3–4 μm, with reportedly promising results.[10–18,21] Lamellar and penetrating keratoplasty are rarely indicated. Though the dystrophy is known to recur in grafts, the disease may be less severe after recurrence.

Vortex Dystrophy

Vortex dystrophy, also known as *cornea verticillata of Fleischer,* is seen as pigmented, whorl-shaped lines in the corneal epithelium.[22,23] This does not represent a corneal dystrophy but rather a degeneration due to toxic keratopathy when the patient is taking phenothiazines, chloroquine, amiodarone, indomethacin, quinacrine hydrochloride, or chlorpromazine.[24,25] It is also seen in the X-linked Fabry's disease. PTK is ineffective. Cessation of medication use results in reduction of corneal findings.

Anterior Crocodile Mosaic Shagreen of Vogt

Clinical Features

First described by Vogt in a 90-year-old woman, this condition is an involutional change which occurs bilaterally and is symmetrical.[26] It is seen as grayish-white polygonal opacities in the basal epithelium and Bowman's layer.[27,28] The intervening areas are clear and resemble crocodile skin. Seen more often anteriorly than posteriorly, it is located centrally. The inheritance is autosomal dominant. Both adult- and juvenile-onset forms occur. The juvenile form is seen with megalocornea, peripheral band keratopathy, and iris malformation.[29] Anterior crocodile shagreen can also be seen after trauma or with band keratopathy.[27] Visual acuity is unaffected, and corneal sensations are normal. Posterior crocodile shagreen occurs due to aging. It is most dense in the posterior stroma and may rarely extend anteriorly to Bowman's layer. A similar anterior mosaic pattern can be seen by flattening the corneal surface and is best demonstrated with applanation tonometry after instillation of fluorescein.

There is deposition of connective tissue between Bowman's layer and the epithelium. Breaks in Bowman's layer can be seen. An irregular sawtooth configuration of collagen lamellae appears on electron microscopy.[30]

Treatment

Generally, no treatment is required for either anterior or posterior shagreen. In a few cases, PTK and keratoplasty may be indicated.

Epithelial Basement Membrane Dystrophy

The underlying pathology in all anterior corneal dystrophies associated with recurrent erosions is abnormal epithelial–basement membrane adhesion complex re-formation. The primary alteration in this complex could be due to either dystrophic epithelium, epithelial basement membrane multilamination, or dystrophic Bowman's layer.

This dystrophy is probably the most frequently encountered anterior corneal dystrophy in clinical practice.[10,31–34] It is also known as *fingerprint/map/dot dystrophy, Cogan's microcystic dystrophy, fingerprint dystrophy, anterior basement membrane dystrophy,* and *dystrophic recurrent erosion.* It is bilateral, with no definite hereditary pattern. Autosomal dominant forms have been encountered.[35–37] Changes morphologically similar to fingerprint/map/dot dystrophy are seen in about 76% of the normal population above the age of 50 years.[38] There is an overlap of epithelial changes between normal individuals following trauma and those with anterior corneal dystrophies. This overlap may reflect the limited ways that the anterior cornea can respond to insults. The term *fingerprint/map/dot dystrophy* aptly describes the clinical appearance of the disorder and represents the historical sequence of recognition of its individual clinical components. Cogan first described microcystic dystrophy of the cornea in five unrelated women who had bilateral grayish-white spheres of varying sizes. Histopathological examination revealed cysts and an abnormal basement membrane. Map-like dystrophy was described in 1965 by Guerry,[39] who had earlier reported fingerprint lines.[40]

Clinical Features

Cogan's dystrophy affects mainly patients above 30 years of age. However, it can occur as early as 4–8 years of age in familial cases. In traumatic recurrent erosions, symptoms occur for a few years, with spontaneous improvement and no significant loss of residual visual acuity.[41] Maps are geographically circumscribed gray lesions best seen with broad oblique illumination, and are the most common and earliest clinical manifestation of fingerprint/map/dot dystrophy (Figure 6.1). Dots are cysts seen in close proximity to the map-like areas. Best seen on retroillumination, they are variable in size, as opposed to the homogeneous cysts of Meesmann's dystrophy (Figures 6.1 and 6.2). The superficial microcysts may stain positively with fluorescein, and the deeper ones may show negative staining indicating unruptured microcysts. Fingerprint lines

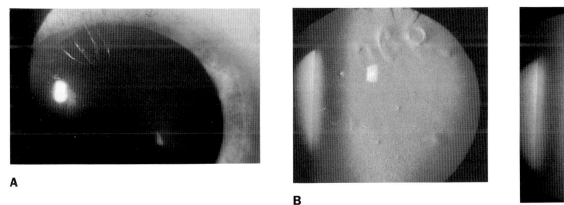
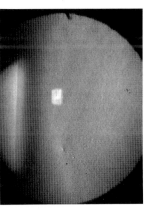
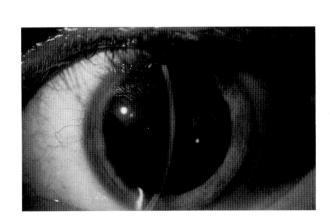

Figure 6.1. Preoperative appearance of a 35-year-old woman with recurrent corneal erosions by direct illumination (*A*) and by retroillumination (*B*). *C, D, E:* Appearance 1, 3, and 12 months following PTK, respectively. There was no recurrence of erosion.

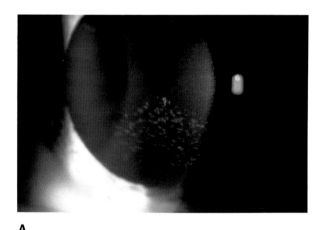 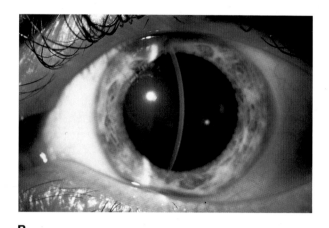

Figure 6.2. A 35-year-old man with epithelial basement membrane dystrophy and persistent recurrent corneal erosion in the left eye after anterior stromal puncture. *A:* Preoperative clinical appearance of the cornea with 20/40 visual acuity. *B:* Postoperative appearance 12 months after PTK with similar visual acuity.

are the features least frequently encountered. They are branching refractile lines with club-shaped terminations and are best seen on retroillumination.[35] They predominate in the central and midperipheral cornea. Combinations of maps and dots are seen most often. Rarely, fingerprints are seen along with maps and dots; dots alone are never seen. Visual acuity is minimally affected. Significant visual loss may occur due to irregular astigmatism.

Histopathology and Pathogenesis

The basic pathology in fingerprint/map/dot dystrophy is abnormal production of basement membrane which may be multilaminated or discontinuous. This may explain the whole spectrum of clinical manifestations. Erosions occur due to absence of hemidesmosomal connections of the basal epithelial cells in areas of abnormal basement membrane. Maps are formed by multilaminar basement membrane and by extension of projections of abnormal basement membrane into the epithelium.[42] Dots are pseudocysts filled with cytoplasmic and nuclear debris and lipid. Cysts form in areas with intraepithelial extensions of aberrant basement membrane where epithelial cells beneath the aberrant basement membrane become vacuolated and liquefied. The wall of the pseudocyst has a corrugated appearance on electron microscopy due to projection of the villous processes of the surrounding epithelial cells into the pseudocyst. This may explain why the cysts vary in size and shape. Rupture of these cysts causes recurrent erosions. Linear projections of fibrillogranular material into the epithelium with thickening of the basement membrane may be the basis of the fingerprints seen on slit lamp examination.

Treatment

In the early stages, treatment consists of symptomatic management of recurrent erosions, including the use of hypertonic saline, patching, and therapeutic soft contact lenses. Epithelial debridement is successful in moderately severe cases.[43] In recalcitrant cases, superficial keratectomy and a soft contact lens may relieve symptoms and reduce recurrences.[44] Anterior stromal puncture (Figure 6.2) may also decrease recurrences.[45] Visual symptoms due to epithelial irregularity can be treated with a hard contact lens, which may also decrease the severity of basement membrane changes. There are several reports of treatment of recurrent erosions with excimer laser PTK, with success rates ranging from 74% to 100%.[12–21] Many of the patients had PTK prior to maximal conventional therapy, and their conditions were not truly recalcitrant. The role of PTK in recalcitrant RES remains to be elucidated.

BOWMAN'S LAYER DYSTROPHIES

Reis-Bücklers' Dystrophy

Initially reported by Reis in 1917[46] and later described in detail by Buckers,[47] this dystrophy is an autosomal dominant superficial corneal dystrophy involving Bowman's layer.[46,47] Two other variants of this dystrophy have been described, namely, Grayson-Wilbrandt and Thiel-Behnke's dystrophies.

Clinical Features
Reis-Bücklers' dystrophy manifests in the first decade of life with recurrent attacks of ocular irritation, photophobia, and watering. These attacks are due to recurrent erosions which decrease in frequency with time as the Bowman's layer is progressively replaced with scar tissue.[46–49] Attacks typically occur four to five times a year. The condition stabilizes after the third decade of life, at which time visual acuity becomes affected. The diffuse opacification and irregular surface contribute to the decreased visual acuity.

Although Reis' original description was that of an annular dystrophy, lesions are usually located centrally and in the midperipheral cornea. The peripheral cornea is spared, and the intervening areas are essentially normal. However, diffuse haze can be seen spreading to the limbus by retroillumination. There is gradual fine reticular opacification of the anterior cornea in the early stages. Superficial gray-white opacities develop at the level of Bowman's layer. They are linear, geographic, honeycombed, and ring-like and cause progressive clouding.[46,49,50] They are best seen with broad oblique illumination. Corneal sensations are decreased, and prominent corneal nerves may be seen.[49,51–53]

Etiology
The structural alterations in Reis-Bücklers' dystrophy may be due to (1) anomalous fibrous tissue production by anterior stromal keratocytes, causing destruction of Bowman's layer; (2) abnormal basal epithelium causing activation of anterior stromal keratocytes, leading to scar formation and secondary destruction of Bowman's layer; or (3) primary dystrophy of the Bowman's layer, causing secondary alterations in the epithelium and stroma. Immunolocalization of laminin and bullous pemphigoid antigen favors the epithelium as the primary cause of this dystrophy.[54] The fact that basement membrane collagens have been localized in the anterior stroma further implicates the epithelium in the pathogenesis of this dystrophy.

Pathology
The histopathological alterations are mainly concentrated in, but not limited to, Bowman's layer.

Bowman's layer is replaced with a fibrocellular material and projects into the epithelium.[56–60] The basal epithelial layers degenerate, and the basement membrane breaks down and is absent in certain areas. The posterior epithelial layer has a sawtooth configuration. In areas where Bowman's layer is intact, the fibrocellular material is accumulated between the epithelium and Bowman's layer.

The fibrocellular layer consists of large collagen fibrils (diameter, 250 to 400 Å) with a regular periodicity interspersed with short, curly filaments with a diameter of 100 Å. There is disorganization of the epithelial–basement membrane adhesion structures, with loss of hemidesmosomes.[61] The basal epithelium shows degenerative changes. There is cytoplasmic vacuolization, mitochondrial swelling, and clumping of nuclear chromatin. Occasionally, the epithelial basement membrane of the bulbar conjunctiva may show reduplication.[62]

Treatment
The fact that the epithelial surface is usually smooth and regular, and that Bowman's layer is scarred and irregular, indicates that PTK may be ideal to correct the visual blurring associated with Reis-Bücklers' dystrophy. In the early stages, recurrent erosions can be treated conventionally with hypertonic saline, patching, or soft contact lenses. Epithelial debridement can be done in cases with recalcitrant erosions as a temporary measure. PTK has been shown to be very successful[18,19,63–69] for the treatment of Reis-Bücklers' dystrophy. Following epithelial debridement or transepithelial ablation, PTK is performed to a depth not usually exceeding 20 μm, with good success. Lamellar keratoplasty and penetrating keratoplasty have been used when visual acuity is seriously affected or the symptoms are debilitating.[48,49,51,70] Recurrences are known to occur after both keratectomy and keratoplasty.[60,71,72] Accordingly, PTK may be the preferred procedure for any patient who is considered for keratoplasty.

Anterior Membrane Dystrophy of Grayson-Wilbrandt

This dystrophy is a variant of Reis-Bückler's dystrophy.[73] Differentiating features include the following: (1) the onset of this disease is at age 10 to 11, slightly later than that of Reis-Bücklers' dystrophy; (2) recurrent erosions are infrequent; (3) visual acuity is not severely affected; and (4) slit lamp examination reveals gray-white mound-like opacities at the level of Bowman's layer projecting into the epithelium. The intervening areas are clear, and the peripheral cornea is spared. Corneal sensation is normal. Grayson-Wilbrandt dystrophy has also been described in a Japanese family whose members had bilateral ring-shaped corneal opacities since adolescence.[74] Histopathological examination reveals PAS-positive material at the level of the basement membrane, especially in places where Bowman's layer is absent. In one patient, ultrastructural examination revealed accumulation of fibrocellular material above an intact Bowman's membrane, with thickening of the basement membrane.[75] Treatment options are similar to those of Reis-Bücklers' dystrophy.

Honeycomb Dystrophy of Thiel and Behnke

This is another variant of Reis-Bücklers' dystrophy. It is a bilateral, autosomal dominant disorder with onset in early childhood.[76] Differentiating features include (1) progressive recurrent erosions which cease later in life followed by a decrease in visual acuity; (2) characteristic honeycomb-like opacification which develops in the central cornea subepithelially; and (3) histological abnormalities in the epithelial basement membrane. The peripheral cornea remains unaffected. The corneal surface is usually smooth between erosion episodes, and the corneal sensations are normal.

Histopathologically, fibrillogranular deposits are seen subepithelially, with projections into the overlying epithelium. The epithelial basement membrane is thickened, split, or duplicated. These findings were also seen in a patient with a family history of Reis-Bücklers' dystrophy, but in the latter condition, Bowman's layer is primarily affected.[77] The treatment of Thiel-Benke dystrophy is similar to that of Reis-Bücklers' dystrophy.

Subepithelial Mucinous Corneal Dystrophy

This dystrophy has some clinical similarities to Grayson-Wilbrandt dystrophy but differs from it histologically. Differentiating features of this dystrophy include (1) late onset (age range, 45 to 78 years); (2) recurrent erosions in early childhood, with progressively decreasing visual acuity ranging from 20/25 to 20/400; (3) diffuse, bilateral, homogeneous subepithelial haze, most dense centrally and fading toward the limbus; and (4) histological evidence of mucinous deposits anterior to Bowman's layer.[78] In some patients there are irregularly shaped, dense, gray-white subepithelial patches in the central and paracentral parts of the cornea, some of which are elevated and distort the anterior corneal contour. The epithelium is intact, and the cornea is of normal thickness. The remaining cornea is unaffected.

Light microscopic examination reveals degeneration of the epithelium without evidence of pseudocysts. A homogeneous eosinophilic layer is noted anterior to Bowman's layer. There is thinning of the overlying epithelium, and the deposits are elevated above the level of the basement membrane. This mucinous material shows a positive PAS reaction and stains with Masson trichome and Alcian blue stains, a differentiating feature from the Grayson-Wilbrandt variant of Reis-Bücklers' dystrophy. Congo red and hyalin staining are negative. Electron microscopy reveals irregular deposits of a fine fibrillar material consistent with proteoglycans, the main component of connective tissue mucins. The epithelial adhesion structures are absent in places where the basement membrane is disrupted. Immunohistochemical staining is positive for chondroitin 4-sulfate and dermatan sulfate.

Treatment includes management of recurrent erosions in the early stages. Penetrating keratoplasty and superficial keratectomy have been performed for visual improvement. As the disease is superficial, excimer laser PTK may be an ideal treatment option.

Local Anterior Mucopolysaccharide Accumulation

This may be another form of Bowman's layer dystrophy similar to subepithelial mucinous corneal dystrophy. It has been reported in two infants who had bilateral corneal clouding with no other manifestations of systemic mucopolysaccharidosis (MPS).[79] Light microscopy revealed intensely staining acid mucopolysaccharide deposits positive for colloidal iron and Alcian blue. The oil-red O stain for lipid was negative. Bowman's layer was thickened and displayed a random orientation of collagen fibrils on electron microscopy. There was no evidence of intracellular vacuoles containing inclusions or extracellular granular material, as seen in systemic MPS. Since this condition has been reported only in infants, the role of PTK is still unknown.

Inherited Band Keratopathy

Band-shaped keratopathy is not usually considered a corneal dystrophy; it is primarily a degeneration seen in ocular diseases like uveitis and phthisis bulbi, as well as in certain systemic diseases. It can also be seen as an inherited trait. The mode of inheritance has not been established. Siblings are more commonly affected, though vertical transmission has been reported.[80-82] The onset can occur at any age. Both childhood and senile forms have been reported. It is similar in appearance to the band-shaped degeneration with yellow or gray opacities in the interpalpebral area due to deposition of calcium in the basal epithelium and Bowman's' layer.

STROMAL DYSTROPHIES

In 1890, Groenouw described a form of corneal degeneration in a family pedigree which was later differentiated into two types by Bückler.[47] He called them *Groenouw I*, now recognized as granular dystrophy, which is autosomal dominant with nodular deposits in the stroma, and *Groenouw II* or macular dystrophy, which is autosomal recessive.

Granular Dystrophy

Clinical Features

Granular dystrophy, or Groenouw I dystrophy, is an autosomal dominant stromal dystrophy with complete penetrance linked to chromosome 5q.[83-86] Onset is in the first or second decade of life. Bilateral, symmetrical, slowly progressive, well-demarcated opacities with intervening clear stromal regions are usually observed. The epithelial surface is usually smooth. The opacities coalesce over time and involve deeper layers of the stroma but never reach the limbus (Figure 6.3).

Three variants of granular dystrophy have been described. In type I the onset is in the first decade of life.[78,87] Characteristic features include (1) progressive snowflake-like opacities; (2) diffuse superficial stromal haze best seen on broad, oblique illumination; (3) fine granularity and punctate opacities seen on indirect illumination and retroillumination; and (4) recurrent erosions. Visual acuity may be severely affected by the fourth decade of life due to stromal opacification.

In type II granular dystrophy, patients are affected in the second decade. It is less severe and slower in progression than type I.[88] Opacities have a breadcrumb-like appearance, and patients do not suffer recurrent erosions.

Rodrigues et al. described a third variant, with an onset in infancy. It is more superficial and causes

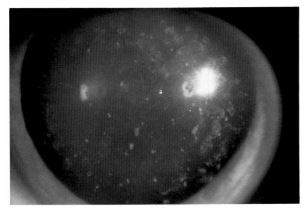
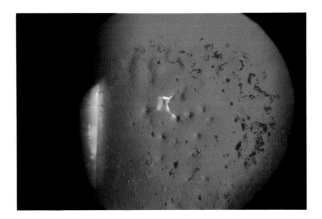

Figure 6.3. Clinical appearance of a patient with granular dystrophy by direct diffuse illumination (*A*) and by retroillumination (*B*).

painful recurrent erosions which require earlier surgical intervention. There are numerous deposits in Bowman's layer, and this layer is destroyed. Classic granular dystrophy is present in parents and/or siblings.[89,90]

Granular dystrophy should be differentiated from paraproteinemic crystalline keratopathy, which may be identical biomicroscopically.[91] Patients with granular dystrophy show a normal immunoglobulin pattern on electrophoresis.

Pathology

The stroma contains hyaline eosinophilic deposits which stain intensely with Masson trichrome, weakly with PAS stain, and negatively with Verhoeff stain. Congo red may stain the peripheral portions of the deposits.[92,93] The epithelial basement membrane and Bowman's membrane may be thin or absent in places.[94]

Ultrastructurally, electron-dense, rod-like deposits (100–500 nm wide) are seen in the subepithelial region within and between keratocytes.[95,96] Tubular microfibrils are also seen. Histochemically, the deposits contain tyrosine, tryptophan and sulfur-containing amino acids.[92,94] Moller et al. demonstrated both kappa and lambda light chains of immunoglobulin G.[97] The exact source of these deposits is unknown. It is speculated that either the keratocytes or the epithelium, or both, are responsible for their production.[98] Those in favor of the epithelial theory believe that a keratocyte-related dystrophy might recur peripherally within the deeper donor tissue after penetrating keratoplasty.

Also, the time taken to recur centrally would be expected to vary proportionally with the graft size, and recurrence should be sooner in a lamellar graft. These facts, however, hold true for macular dystrophy but not for granular dystrophy.

Degenerative changes are apparent in keratocytes with cytoplasmic vacuolization and dilatation of endoplasmic reticulum.[99] In recurrent cases, electron-dense deposits are seen in the epithelium. These are crystalline granules which are either rod- or trapezoid-shaped. The smaller granules are surrounded by 15-nm particles, whereas the larger granules are membrane bound.[100–102]

Treatment

Intervening clear areas between the opacities allow unobstructed vision in the early stages of the disease. With progression of the disease, visual acuity is affected and superficial keratectomy is warranted if the deposits are superficial.[103] Superficial variant treatment with the excimer laser may offer fine tuning of the surgical depth and is thus an attractive alternative to lamellar keratectomy. The goal of PTK is to remove the anterior corneal opacity and smooth the anterior refractive surface. PTK is performed after removal of the epithelium, either mechanically or ablated with the excimer laser. If the epithelial surface is rough, a smoothing agent or modulator fluid is used to create a more regular surface that is then treated. Success rates ranging from 66% to 100% have been reported[104] (Figure 6.4). Penetrating keratoplasty is required in the more advanced cases. Recurrences after keratectomy

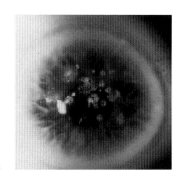
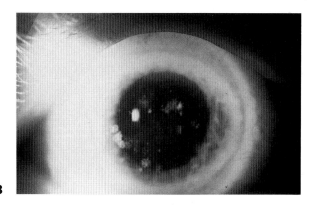

Figure 6.4. *A:* Clinical appearance of a patient with granular dystrophy. *B:* Appearance 1 month after PTK. Visual acuity of 20/20 was achieved despite residual deposits of hyaline material in the visual axis.

have been reported as early as 1 year after surgery and are more frequent in the superficial variant.[105,106] It is believed that patients with homozygous disease present earlier, in a more severe form, and have earlier recurrences following surgical intervention.[107] Recurrences are seen as a diffuse haze in the peripheral graft or as granular lesions in the stroma. The diffuse haze occurs due to growth of fibrous tissue between the epithelium and Bowman's layer. It can be stripped by superficial keratectomy (Figure 6.5).

Lattice Dystrophy

Lattice dystrophy, or Biber-Haab-Dimer dystrophy, has been subclassified into three distinct forms. Type I (the classic form) and type II (the systemic form) are autosomal dominant. Type III is autosomal recessive. The disease may affect a few family members, while other members may show only recurrent erosions with minimal stromal opacification[108] (Figure 6.6). It is bilaterally symmetrical, though unilateral cases have been reported.[109–111] Unilateral cases have later onset and a less severe clinical course.

Clinical Features

Type I lattice dystrophy usually appears in the first decade of life.[112] In the early stages it appears as irregular lines and dots in the anterior axial stroma or as a diffuse haze in the center. The faint central haze in the center becomes more dense with time. This central disc-shaped opacity eventually causes reduced visual acuity and may even obscure the underlying lattice pattern (Figure 6.7). Lattice lines may extend to the periphery. They branch dichotomously near their central terminations and overlap one another at various stromal levels. Small dots and opacities may be present between the lines. As the opacities coalesce, a diffuse haze appears involving the anterior and middle stroma. At this stage the lattice dystrophy may be difficult to dis-

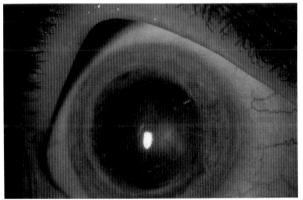

A

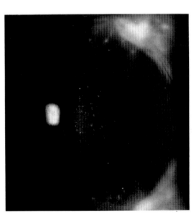

C

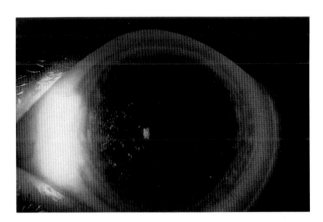

B

Figure 6.5. PTK for recurrent granular dystrophy in a 56-year-old woman. *A:* Clinical appearance of recurrent granular dystrophy showing superficial, mild, diffuse opacification resulting in 20/70 visual acuity. *B:* Unoperated contralateral eye showing typical granular dystrophy and 20/100 visual acuity. Note the difference in clinical appearance between the two eyes. *C:* Appearance of the cornea with recurrent granular dystrophy 1 year following PTK, with improvement of vision to 20/40.

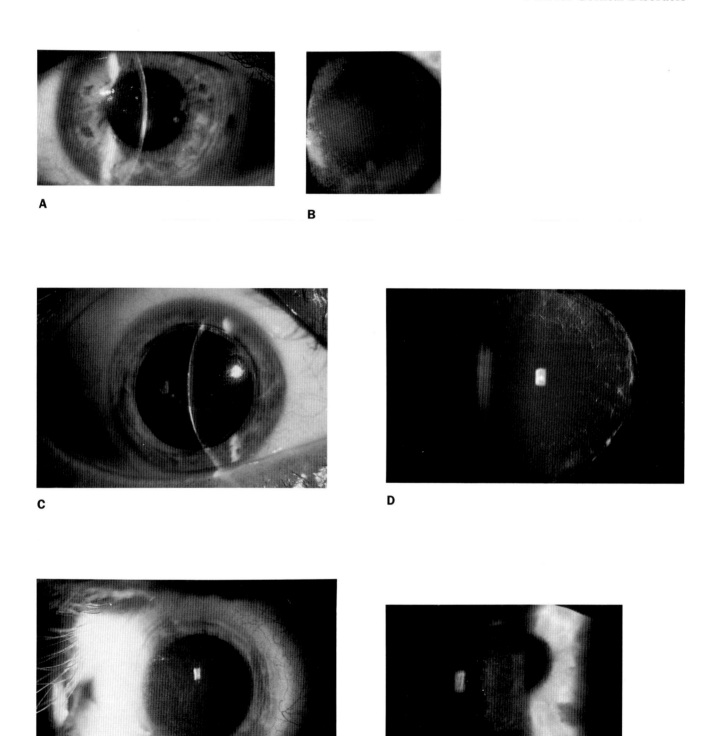

Figure 6.6. Clinical appearance of the corneas of five brothers with lattice dystrophy who underwent PTK. *A, B:* Member 1: appearances of the cornea preoperatively and 3 months after PTK, respectively. *C, D:* Member 2: before PTK and 6 months after PTK, respectively. *E, F:* Member 3: before PTK and 3 months after PTK, respectively.

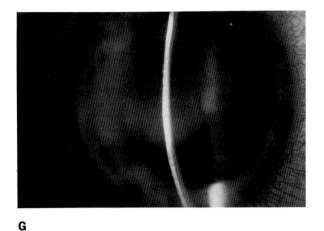
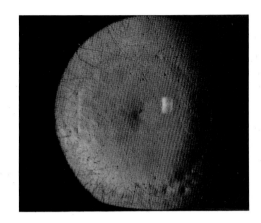
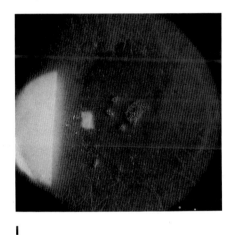

Figure 6.6. *G:* Member 4: 5 months postoperatively. *H, I:* Member 5: 5 and 14 months after PTK, respectively.

tinguish from granular and macular dystrophies, but careful examination usually reveals the typical branching lattice lines. In advanced cases the lattice lines fluoresce under cobalt blue slit lamp illumination. Pseudofilaments may form due to the linear configuration of dots. Minute gray opacities in the superficial stroma may resemble granular or macular dystrophy in the early stages, but their refractile quality in lattice distinguishes them from other stromal dystrophies.

Epithelial erosions occur in the second decade and are the cause of irregular astigmatism and decreased visual acuity. They diminish in frequency after the fourth decade. Vascularization occurs rarely.

An association between vestibulocochleopathy and lattice corneal dystrophy has been reported in a case of nonfamilial amyloidosis.[113] Coexistence of progressive external ophthalmoplegia and corneal lattice dystrophy has also been reported.[114]

Lattice type II (Meretoja) corneal dystrophy was originally reported in 1969.[115-117] It is associated with systemic amyloidosis and is inherited as an autosomal dominant disorder. It is also known as *familial amyloid polyneuropathy (FAP) type IV*. The corneal changes appear in the third to fourth decade of life. Systemic involvement of the skin, peripheral nerves, and cranial nerves occurs later. This dystrophy is less severe than lattice type I corneal dystrophy, with fewer dots and lattice lines which are radial, thicker, and extend to the limbus. Erosions are less frequent, and vision is unaffected until the seventh decade of life.[118]

Lattice type III corneal dystrophy was first reported in two families in Japan.[119,120] It manifests as thick, translucent lattice lines and diffuse subep-

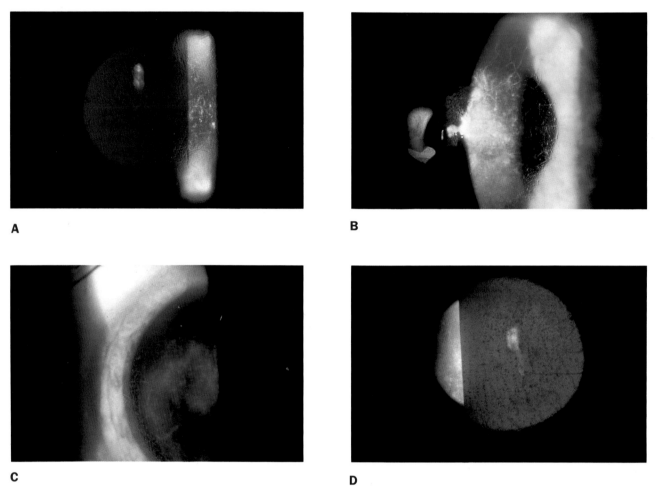

Figure 6.7. A 45-year-old man with a history of progressive lattice dystrophy who underwent PTK. *A–C:* Preoperative appearance of the cornea showing progression of dystrophy over 12 years, resulting in 20/1000 visual acuity and a 2+ mean haze score. *D:* Retroillumination of the eye 12 months after PTK showing residual lattice dystrophy. Visual acuity improved to 20/200, with a mean haze score of 0.

ithelial opacities. Disease onset occurs after age 40, and vision is not affected until 70 years of age, but the condition is not associated with systemic amyloidosis. Corneal erosions are not seen.[119] A variant of this dystrophy, named *lattice type III A,* has a similar clinical appearance. However, its recurrent erosions and its location in whites differentiate it from lattice type III.

Pathology

The epithelial layer is irregular. Descemet's membrane and endothelium are unaffected. Involvement of the epithelial or Bowman's layer can be confirmed by polarized microscopy. The presence of amyloid deposits in lattice dystrophy has been confirmed by immunofluorescence, Congo red, PAS, and Massons trichrome staining.[120] The deposits exhibit dichroism (alternate red and green color when viewed through a rotating polarizing filter) and manifest green birefringence.[121–123] In type II lattice corneal dystrophy, amyloid is deposited systemically, and amyloid deposits replace corneal nerves with diffuse gelsoin exclusive immunolocation.[124,125]

Ultrastructurally, the epithelium is of variable thickness. There is degeneration of the basal epithelium. The basement membrane is thickened and continuous, without normal hemidesmosomes.[126,127] Bowman's layer and superficial stroma contain amyloid structures, collagen fibrils, and fibroblasts. Amyloid is a complex of chondroitin, sulfuric acid, and protein and is found extracellularly. The amyloid seen in lattice dystrophy types I and III is distinct from that found in systemic amyloidosis or in primary or secondary amyloid degeneration of the cornea. Lattice dystrophy types I and III contain both protein AA and protein AP, while lattice dystrophy type II contains either protein AA or protein AP.[119,128] Fluorescent staining for immune deposits like kappa chains, lambda chains, immunoglobulin G (IgG), IgM, and IgA seen in primary or secondary systemic amyloidosis is negative.[129,130] It is believed that sequestration of glycoprotein in the corneal stroma from plasma membranes of epithelial cells may stimulate amyloid deposition.[131]

Treatment

Management of lattice dystrophy in the early stages includes treatment of recurrent erosions. PTK with the excimer laser is emerging as an attractive alternative to treat superficial corneal opacities and surface irregularities. The goal is to obviate more invasive procedures such as lamellar and penetrating keratoplasty. PTK has been reported to be very successful for the treatment of lattice dystrophy.[104] Penetrating keratoplasty is indicated when there is a decrease in visual acuity or, rarely, when the recurrent erosions become incapacitating. It is generally required in the third or fourth decade of life. Recurrence of lattice dystrophy in the graft is seen as early as 3 years postoperatively. Recurrences are more frequent in lattice dystrophy than in granular and macular dystrophies, and are seen as superficial dot-like, filamentous subepithelial opacities or as a diffuse stromal haze[127] (Figure 6.8).

Avellino Corneal Dystrophy

Folberg et al. reported three families with histological features of both granular and lattice dystrophies.[133] As all of these patients traced their origins to Avellino, Italy, the condition was later named *Avellino corneal dystrophy* by Holland and colleagues.[134] Earlier reports suggested that there may be a close relationship between lattice type 1 and granular dystrophy amyloid deposits seen in patients with clinically typical granular dystrophy; conversely, granular deposits had been detected in corneal buttons with otherwise typical lattice type 1 dystrophy.[135] Avellino dystrophy is inherited as an autosomal dominant trait with high penetrance and variable expressivity. Stone et al. performed linkage analysis for three clinically distinct stromal corneal dystrophies: lattice, granular, and Avellino.[136] They found that the location is the same on chromosome 5q for all three dystrophies.

Clinical Features

This dystrophy usually appears in the first or second decade of life. Characteristic features include (1) anterior stromal discrete, gray-white granular deposits, (2) middle to posterior stromal lattice lesions, and (3) anterior stromal haze.

The earliest findings are those of discrete granular deposits in the anterior third of the corneal stroma. Granules reach their mature size early. They coalesce to form linear opacities, especially in the inferior cornea. Lattice changes start later in life in the middle and deep stroma and increase in proportion to the age of the patient to involve the entire stroma. The lattice rods are thicker than those in lattice corneal dystrophy type 1. The lattice component of Avellino dystrophy resembles lattice corneal dystrophy type IIIA.[137] The diffuse "ground-glass" stromal haze located between the granular deposits is the latest clinical finding.

Visual acuity is minimally affected and depends upon the predominant location of the deposits, which may be either central or peripheral. Patients complain of glare and decreased night vision. Recurrent erosions occur in the third and fourth decades of life secondary to the presence of superficial granular deposits.[138] Recurrent erosions are more common in patients with Avellino corneal dystrophy than in those with typical granular dystrophy.

Pathology

Granular deposits demonstrate a hyaline appearance by hematoxylin-eosin staining and stain bright red with Masson's trichrome. Coexisting are numerous fusiform stromal deposits of amyloid that stain with Congo red and demonstrate birefringence and dichroism with cross-polarization, which is typical of lattice dystrophy.[139,140] The superficial granular deposits are ultrastructurally homogeneous and electron dense. Microfilaments suggestive of amyloid can be seen in some of these deposits.

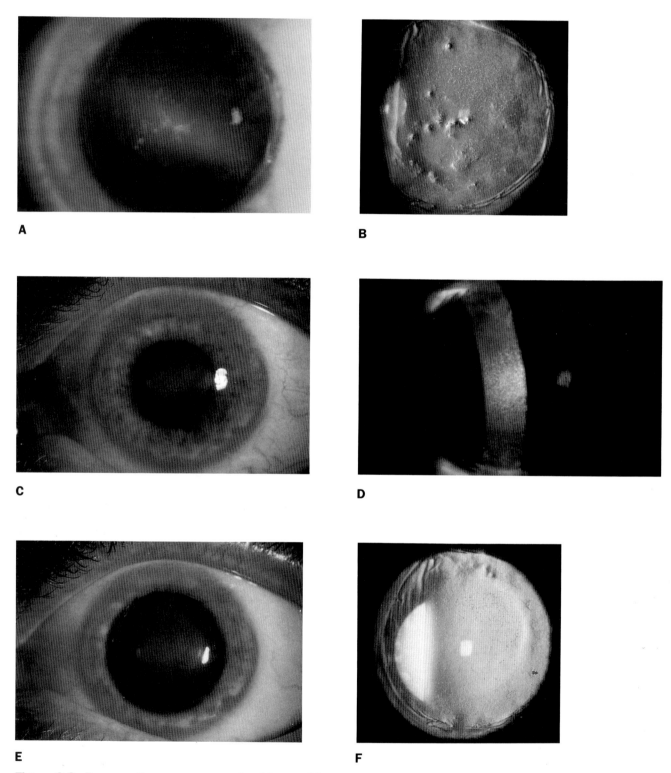

Figure 6.8. Preoperative appearance of a 48-year-old man with recurrent lattice dystrophy in a graft by direct illumination (*A*) and by retroillumination (*B*). *C:* Slit lamp photograph of the eye 1 month after PTK. *D:* Appearance 2 months after PTK. *E, F:* Appearance 3 months after PTK by direct illumination and retroillumination respectively.

Anterior Corneal Dystrophies: Clinical Features

Treatment

Treatment is conservative and includes hypertonic saline and bandage contact lenses for recurrent erosions. PTK with the excimer laser has been used to treat corneal erosions and to clear the central cornea of granular and lattice deposits.[141] Treatment for decreased visual acuity includes penetrating keratoplasty and is generally required late in the course of the disease. Recurrences of the granular lesions have been reported as late as 9 years after keratoplasty.[134]

MACULAR DYSTROPHY

Clinical Features

Macular dystrophy is the least common but most severe of the three classic stromal dystrophies. It is transmitted as an autosomal recessive trait. Unlike granular dystrophy, whose corneal opacities have distinct edges, the intervening stroma in macular dystrophy is not clear, and the clouding extends peripherally and into the deeper stroma as the disease progresses. The corneal lesions usually are first noted between 3 and 9 years of age, when a diffuse clouding is seen in the central superficial stroma. The cornea becomes increasingly cloudy as the condition progresses. The opacification involves the entire thickness of the cornea and may extend out to the limbus by the teens. The denser macular opacities, which have ill-defined borders, can protrude anteriorly, resulting in irregularity of the epithelial surface, or posteriorly, causing irregularity, grayness, and a guttate appearance of Descemet's membrane. The opacities can enlarge with time and coalesce, and the patients may suffer from progressive loss of vision, irritation, and photophobia. Recurrent erosions are seen, but they are less frequent than in lattice dystrophy. In most cases, vision is severely impaired by age 20 or 30. The cornea may become thinner in macular dystrophy.[142,143] Heterozygous carriers do not manifest corneal abnormalities.

The primary defect of macular dystrophy is in the synthetic pathway of the keratan sulfate proteoglycans,[144–146] which has incomplete glycosaminoglycan sulfation resulting from decreased activity of α-galactosidase in keratocytes.[147] Macular dystrophy is subdivided into two types. Type 1 is most prevalent in Europe and North America. It is characterized by the absence of antigenic keratan sulfate in the cornea, as well as in the serum, and it may represent a more widespread systemic disorder of keratan sulfate metabolism.[148,149] Using synchrotron x-ray diffraction, Quantock et al.[150] found that the interfibrillar spacing of collagen fibrils was significantly lower in type 1 macular dystrophy than in normal adult human corneas. They suggested that this close packing of collagen fibrils seemed to be responsible for the reduced thickness of the central cornea in macular dystrophy.[150] Type 2 may be more prevalent in Japan, based on a limited study by Santo et al.,[151] and keratan sulfate is present in the cornea and serum.[152,153]

Pathology

Histologically, macular dystrophy is characterized by the accumulation of glycosaminoglycans[154] (acid mucopolysaccharide) between the stromal lamellae, underneath the epithelium, within stromal keratocytes, and within the endothelial cells.[155–159] The glycosaminoglycans stain intensely with alcian blue and colloidal iron, minimally with PAS, and not at all with Masson's trichrome. Bifringence is decreased. Progeoglycan-specific Cuprolinic blue stains have revealed that these proteoglycans accumulate and aggregate in corneas with both type 1 and type 2 macular dystrophy. Degeneration of the basal epithelial cells and focal epithelial thinning are seen over the accumulated material. Bowman's layer is thinned or absent in some areas.

Electron microscopy shows accumulation of mucopclysaccharide within stromal keratocytes.[156,159–162] The keratocytes are distended by numerous intracytoplasmic vacuoles, which appear to be the dilated cisternae of the rough endoplasmic reticulum. Some of these vacuoles are clear, but many contain fibrillar or granular material and occasionally membranous lamellar material. The endothelium contains similar material. The posterior, nonbanded portion of Descemet's membrane is infiltrated by vesicular and granular material deposited by the abnormal endothelium. Quantock et al.[163] found abnormally large collagen fibrils which existed in localized regions, frequently adjacent to the vacuoles, and abnormal proteoglycans in an atypical variant of macular corneal dystrophy.

The accumulated material varies; its staining by different anti–keratan sulfate antibodies differs among patients.[164,165] In some patients, normal keratan sulfate is also absent from the serum and cartilage.[164,166] The keratocytes produce only glycoprotein precursors of keratan sulfate in culture.[145,167,168] In another study, the deposits were found to contain an antigen associated with intermediate-type filament.[169] With the acridine orange technique, compensatory generalized hyperactivity of the lysosomal enzyme system has been demonstrated.[170]

As is typical of an autosomal recessively inherited condition, macular dystrophy presumably results from deficiency of a hydrolytic enzyme (sulfotransferase) and may thus be considered a localized form of MPS.[144] The corneal pathology of macular corneal dystrophy, however, differs from that of the systemic MPSs in several ways. In the systemic MPSs, the abnormal material accumulates in lysosomal vacuoles, whereas in macular dystrophy it accumulates in endoplasmic reticulum. In the systemic MPSs, epithelial involvement is prominent and Descemet's membrane usually is not affected.

Treatment
Excimer laser phototherapeutic keratectomy has been tried to remove the corneal opacities in macular dystrophy; however, the result is not as good as that achieved in other superficial dystrophies or in granular dystrophy.[171] For those whose vision fails to improve with PTK, however, good results are obtained with penetrating keratoplasty. Recurrences can be seen in both lamellar and penetrating grafts, but they usually are delayed for many years.[172,173] Host keratocytes invade the graft and produce abnormal glycosaminoglycans. The periphery of the graft is most severely affected, particularly the superficial and deep layers. The endothelium and Descemet's membrane also are involved.

CENTRAL CRYSTALLINE DYSTROPHY OF SCHNYDER

Clinical Features
Central crystalline dystrophy is a slowly progressive autosomal dominant dystrophy characterized clinically by bilateral central corneal opacities that are sometimes associated with premature corneal arcus and limbal girdle of Vogt.[174-176] Xanthelasma[177] and dyslipidemia can also be seen in some cases. The severity of dyslipidemia, which may[175,178-180] or may not[181-184] be associated with Schnyder dystrophy, usually does not correlate with the extent of opacification or the progression of the disease.[185] A significant number of patients have hyperlipidemia, but the type and severity of the hyperlipidemia vary.[186,187] Within a single family some individuals may have only crystalline dystrophy, others may have hyperlipidemia and crystalline dystrophy, and still others may have only hyperlipidemia. Chondrodystrophy and genu valgum also are associated with Schnyder's dystrophy in some families.[175,186-188]

Schnyder's dystrophy usually occurs in early life and may be seen as soon as 2 months of age but usually goes undetected at that age.[174] Cogenital cases are seen sporadically. The patient's visual acuity usually is not affected, but occasionally it is moderately reduced. The main feature of the dystrophy is the presence of bilateral, axial, ring-shaped or disciform opacities.[175,186,187,189,190] The corneal opacities usually consist of fine, polychromatic, needle-shaped crystals, but in some cases a disciform opacity is present without evident crystals[174] (Figure 6.9). In Weiss's series, only 51% of patients with Schnyder crystalline dystrophy had clinical evidence of corneal crystalline deposits.[191] On rare occasions, clouding extends to the arcus,[192] but usually a clear cornea persists between the central opacification and the surrounding arcus.[193] The yellow-white opacity primarily involves Bowman's layer and the anterior stroma but may extend into the deeper layers.[177] The epithelium is normal and the intervening stroma usually is clear, but punctate white opacities can be scattered in the stroma or the stroma can develop a milky opalescence.[186,187] Corneal sensation is normal. Progression is more frequent in patients with diffuse opacities than in those with crystalline deposits.[185]

Pathology
The main histopathological feature of Schnyder's dystrophy is the presence of phospholipid, cholesterol crystals, noncrystalline cholesterol, cholesterol esters, and neutral fats in the stroma.[177,184,187,189,194] The clinically apparent crystals correspond to cholesterol accumulations, both within keratocytes and extracellularly. Occasionally, similar deposits have been noted in basal epithelial cells.[181] The deposits are most numerous in the anterior stroma, but they can extend posteriorly to Descemet's membrane.[195]

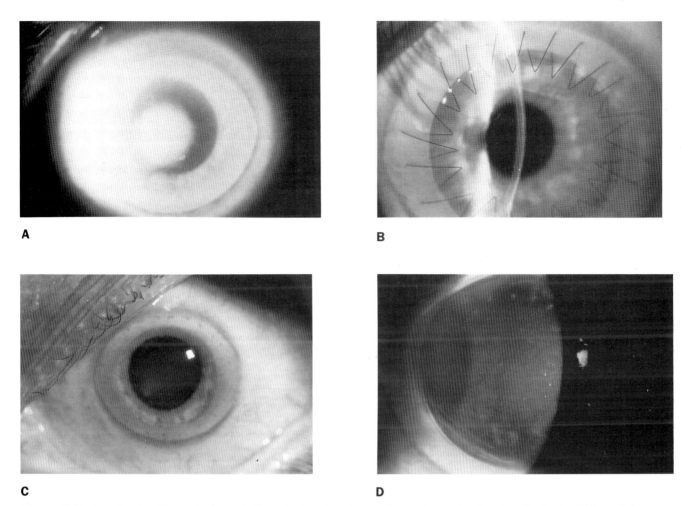

Figure 6.9. A patient with central crystalline dystrophy who underwent penetrating keratoplasty (PK) and, 5 years later, photoastigmatic keratectomy (PAK) for correction of myopia and astigmatism. *A:* Preoperative appearance. *B:* Clinical appearance following PK. *C:* Preoperative appearance before PAK. *D:* Postoperative appearance 3 months after PAK.

Destruction of Bowman's layer and superficial stroma with disorganization of collagen have often been observed.

The pathogenesis of Schnyder's dystrophy, although unclear, is thought to involve a primary disorder of corneal lipid metabolism,[175,177,179,193] the severity of which may be altered by systemic hyperlipidemia. Bums et al.[196] administered radio-labeled cholesterol intravenously to a patient with crystalline dystrophy 2 weeks before keratoplasty. The radioactivity in the cornea was higher than that in the blood, suggesting active deposition of cholesterol in the cornea. The primary abnormality would involve excessive production or diminished breakdown of phospholipids, unesterified cholesterol, or other constituents of cell membranes. This concept can be supported by the findings that keratocytes and their membranes are abnormal in Schnyder dystrophy[175,196] and in the characteristic membrane-bound vacuoles that appear to bud from degenerating keratocytes.[180] Both the vacuoles and keratocytes were noted to have similar trilamellar membranes.[177] Such budding could represent breakdown of unstable cell membranes composed, in part, of excess phospholipid and unesterified cholesterol. Mirshahi et al.[197] showed much lower secretion of plasminogen activators by corneal fibroblasts from patients with Schnyder's dystrophy than that secreted by normal corneal fibroblasts. Since plasminogen activators are involved in extracellular matix remodeling, the deficiency may be responsible for diminished breakdown of cell membranes in corneal dystrophy.

Treatment

In most cases of crystalline dystrophy, the corneal disease requires no treatment. Serum lipid profiles should be obtained, although the severity of a systemic lipid abnormality does not necessarily correlate with the severity of the corneal disease. Elevated serum lipid levels and concomitant cardiovascular disease are associated features in some patients.[175] Efforts directed at visual improvement through such means as reduction of cholesterol intake have proven beneficial in only one case report.[198] Excimer laser PTK can be perform to remove the central superficial opacities, but the result is not well established. Penetrating or lamellar keratoplasty can be performed for visual rehabilitation if PTK fails to improve the vision, but the dystrophy can recur.[175,199] Crystalline deposits recur sooner and in larger amounts in lamellar grafts than in full-thickness grafts.[174] Postoperative changes in the posterior layers of the cornea in Schnyder's crystalline dystrophy may contribute to the comparatively poorer surgical outcomes obtained with lamellar grafts compared to full-thickness grafts.[195]

GELATINOUS DROP-LIKE DYSTROPHY (PRIMARY FAMILIAL AMYLOIDOSIS OF THE CORNEA)

Clinical Features

Gelatinous drop-like dystrophy is a rare familial disorder of the cornea.[200–204] Early investigators in the United States termed the condition *primary familial amyloidosis of the cornea*.[205,206] A similar condition was described in the European literature in 1930.[207] The pattern of inheritance is unclear but is most likely autosomal recessive.[205] Gelatinous drop-like dystrophy, although rare in America and Europe, is much more common in Japan.[208] It accounted for the largest number of keratectomy/keratoplasty specimens in one Japanese study.[209]

The disorder usually appears in the first decade with photophobia, lacrimation, and decreased visual acuity. The corneal lesion is characterized by bilateral, central, raised, multinodular, subepithelial mounds of amyloid. They are white on direct illumination and transparent on retroillumination but can become yellow and milky with time. Flat subepithelial opacities can be seen surrounding the mounds. The deposits are fairly flat and can resemble band-shaped keratopathy early in the dystrophy.[210] However, they increase in number and depth with age. In the late stages of the disease, the cornea can have a diffuse, mulberry-like appearance. Vascularization, if present, is usually minimal, and anterior and posterior cortical lens changes have been reported.[205]

Pathology

The corneal deposits are amyloid both histologically and ultrastructurally.[201–203,205,206,211] Bowman's layer usually is absent, and amyloid is deposited in the basal epithelial cells. There are also fusiform deposits similar to those of lattice dystrophy in the deeper stroma. A flat, more uniform layer of a similar material may surround the nodular masses. Amyloid can also be deposited in the deep stroma.[209] In an unusual case, large amyloid deposits were found to a depth of two-thirds of the cornea, which contained needle-like cholesterol crystals.[209] Small spheroidal deposits showing a positive reaction to Verhoeff iron hematoxylin were demonstrated at the periphery of the cornea in some cases. The type of corneal amyloid deposits containing protein AP but not protein AA or immunoglobulins may be different from that found in lattice dystrophy.[212]

The fact that perilimbal conjunctiva showed amyloid deposits in the stroma[209] supported the idea that the epithelial cells, more precisely the limbal cells, are involved in the synthesis of amyloid fibrils and might have a role in the recurrence process. This also would explain the success observed when lamellar keratoplasty is combined with keratoepithelioplasty around the limbal area after a 360° peritomy.[213] However, the finding of stromal deposits suggests that keratocytes may participate in some way in the pathogenesis of this disease. Weber and Babel[203] suggested the possibility that gelatinous drop-like dystrophy and lattice dystrophy might be two facets of the same basic disorder, with the former representing the epithelial expression and the latter the stromal expression.

Treatment

Excimer laser PTK can be attempted to remove superficial lesions. Other alternatives for deeper deposits include either superficial keratectomy or keratoplasty. Recurrence is seen frequently after penetrating keratoplasty.[209] In fact, recurrences in a graft accounted for 29.6% of the patients undergoing corneal surgery in one Japanese series.[209]

REFERENCES

1. Pameijer JK: Ueber eine Fremdartige FamiliEre Oberfl Echliche Hornhautver Enderung. *Klin Monatsbl Augenheilkd* 1935;95:516.
2. Meesmann A, Wilke F: Klinische und Anatomische Untersuchungen Ueber eine BisherUnbekannte, Dominant Vererbte Epitheldystrophie der Hornhaut. *Klin Monatsbl Augenheilkd* 1939;103:361.
3. Stocker FW, Holt LB: Rare form of hereditary epithelial dystrophy. *Arch Ophthalmol* 1955;53:536.
4. Meesmann A: Ueber eine Bisher Nicht Beschriebene Dominant Vererbte Dystrophia Epithelialis Corneae. *Ber Dtsch Ophthalmol Ges* 1938;52:154.
5. Kuwabara R, Ciccarelli EC: Meesmann's corneal dystrophy. *Arch Ophthalmol* 1964;71:676.
6. Snyder WB: Hereditary epithelial corneal dystrophy. *Am J Ophthalmol* 1963;55:56.
7. Tremblay M, Dube L: Meesmann's corneal dystrophy: Ultrastructural features. *Can J Ophthalmol* 1982;17:24.
8. Burns RP: Meesmann's corneal dystrophy. *Trans Am Ophthalmol Soc* 1968;66:531.
9. Fine BS, Yanoff M, Pitts E, Slaughter FD, et al: Meesmann's epithelial dystrophy of the cornea. *Am J Ophthalmol* 1977;83:633.
10. Cogen DG, Kuwabara T, Donaldson DD, Marshall D: Microcystic dystrophy of the corneal epithelium. *Trans Am Ophthalmol Soc* 1964;62:213.
11. Nakaniski I, Brown SI: Ultrastructure of the epithelial dystrophy of Meesmann. *Arch Ophthalmol* 1975;93:259.
12. Ohman L, Fagerholm P, Tengroth B: Treatment of recurrent corneal erosions with the excimer laser. *Acta Ophthalmol* 1994;72:461.
13. Dausch D, Landesz M, Klein R, Schroder E: Phototherapeutic keratectomy in recurrent corneal erosion. *Refract Corneal Surg* 1993;9:419.
14. Fagerholm P, Fitzsimmons TD, Orndahl M, Ohman L, Tengroth B: Phototherapeutic keratectomy: Long-term results in 166 eyes. *Refract Corneal Surg* 1993;9(Suppl 2):76.
15. Foster W, Grewe S, Atzler U, Lunecke C, Busse H: Phototherapeutic keratectomy in corneal diseases. *Refract Corneal Surg* 1993;9(Suppl 2):85.
16. Rapuano CJ, Laibson PR: Excimer laser phototherapeutic keratectomy. *CLAO J* 1993;19:235.
17. John ME, Karr Van D, Noblitt RL, Boleyn KL: Excimer laser phototherapeutic keratectomy for treatment of recurrent corneal erosion. *J Cataract Refract Surg* 1994;20:179.
18. Sher NA, Bowers RA, Zabel RW, et al: Clinical use of the 193-nm excimer laser in the treatment of corneal scars. *Arch Ophthalmol* 1991;109:491.
19. Hersh PS, Spinak A, Garrana R, Mayers M: Phototherapeutic keratectomy: Strategies and results in 12 eyes. *Refract Corneal Surg* 1993;9(Suppl 2):90.
20. Niesen U, Thomann U, Schipper I: Phototherapeutic keratectomy [in German] *Klin Monatsbl Augenheilkd* 1994;205:187.
21. Poirier C, Coulan P, Williamson W, Mortemousque B, Verin P: Results of therapeutic photo-keratectomy using the excimer laser. Apropos of 12 cases (review [in French]) *J Fr Ophthalmol* 1994;17:262.
22. Johnson BL, Brown SI, Zaidman GW: A light and electron microscopic study of recurrent granular dystrophy of the cornea. *Am J Ophthalmol* 1981;92:49.
23. Ruusuvaaara P, Setala K, Tarkkanen A: Granular corneal dystrophy with early stromal manifestation. *Acta Ophthalmol (Copenh)* 1990;68:525.
24. Witschel H, Sundmacher R: Bilateral recurrence of granular dystrophy in the grafts. *Graefes Arch Clin Exp Ophthalmol* 1979;209:179.
25. Moller HU, Ehlers N: Early treatment of granular dystrophy (Groenouw type 1). *Acta Ophthalmol (Copenh)* 1985;63:597.
26. Wogt A: *Textbook and Atlas of Slit Lamp Microscopy of the Living Eye.* Bonn, Wayenborgh Editions, 1981.
27. Tripathi RC, Bron AJ: Secondary anterior crocodile shagreen of Vogt. *Br J Ophthalmol* 1975;59:5.
28. Pouliquen Y, Dhermy P, Presles D, Tollard MF: De generescence en Chagrin de crocodile de Vogt ou denenerescence en mosaique de Valerio. *Arch Ophthalmol (Paris)* 1976;36:395.
29. Carenini BB, Brogliatti B: Laser treatment of glaucoma (review [in French]). *Cesk Ophthalmol* 1988;44:377.
30. Krachmer JH, Dubord PJ, Rodrigues MM, Mannis

MJ: Corneal posterior crocodile shagreen and polymorphic amyloid degeneration. *Arch Ophthalmol* 1983;101:54.
31. Cogan DG, Kuwabara T, Donaldson DD, Collins E, et al: Microcystic dystrophy of the cornea. *Arch Ophthalmol* 1974;92:470.
32. King RG Jr, Geeraets R: Cogan-Guerry microcystic corneal epithelial dystrophy. *Med Coll Va Q* 1972; 8:241.
33. Trobe JD, Laibson PR: Dystrophic changes in the anterior cornea. *Arch Ophthalmol* 1972;87:378.
34. Wolter JR, Fralick FB: Microcystic dystrophy of the corneal epithelium. *Arch Ophthalmol* 1966;75:380.
35. Laibson PR, Krachmer JH: Familial occurrence of dot, map, and fingerprint dystrophy of the cornea. *Invest Ophthalmol* 1975;14:397.
36. Franceschetti A: Hereditdre Rezidivierende Erosion der Hornhaut. *Z Augenheilkd* 1928;66:309.
37. Bron AJ, Buriless SEP: Inherited recurrent corneal erosion. *Trans Ophthalmol Soc UK* 1981;101:239.
38. Werblin TP, Hirst LW, Stark WJ, Maumenee IH: Prevalence of map-dot-fingerprint change in the cornea. *Br J Ophthalmol* 1980;65:401.
39. Guerry DUP III: Observations on Cogan's microcystic dystrophy of the corneal epithelium. *Trans Am Ophthalmol Soc* 1965;63:320.
40. Guerry DUP III: Fingerprint-like lines in the cornea. *Am J Ophthalmol* 1950;33:724.
41. Waring GO, Rodriquez MM, Laibson PR: Corneal dystrophies. 1. Dystrophies of the epithelium, Bowman's layer, and stroma. *Surv Ophthalmol* 1978; 23:71.
42. Broderick JD, Dark AJ, Peace GW: Fingerprint dystrophy of the cornea. *Arch Ophthalmol* 1974;92:483.
43. Nirankari VS, Rodrigues MM, Jarmarwala MG, Rajagapalan S, et al: An unusual case of epithelial basement membrane dystrophy. *Am J Ophthalmol* 1989;107:552.
44. Buxton JN, Fox ML: Superficial epithelial keratectomy in the treatment of epithelial basement membrane dystrophy. *Arch Ophthalmol* 1983;101:392.
45. Katsev DA, Kincaid MC, Fouraker BD, Dresner MS, Schanzlin DJ: Recurrent corneal erosion: Pathology of corneal puncture. *Cornea* 1991; 10:418.
46. Reis W: Familidre, Fleckige Hornhautentartung. *Dtsch Med Wochenschr* 1917;43:575.
47. Bücklers M. Über eine weitere familiare Hornhautdystrophie. *Klin Monatsbl Augenheilkd* 1949;114:386.
48. Rice NSC, Ashton N, Jay B, Black RK: Reis-Bucklers' dystrophy. *Br J Ophthalmol* 1968;52:577.
49. Jones ST, Stauffer LH: Reis-Bücklers' corneal dystrophy. *Trans Am Acad Ophthalmol Otolaryngol* 1970;74:417.
50. Paufique L, Bonnet M: La dystrophie corneenne heredo-familiale de Reis-Bucklers. *Ann Oculist* 1966;199:14–37.
51. Hall P: Reis-Bucklers' dystrophy. *Arch Ophthalmol* 1974;91:170.
52. Kaufman HG, Clowe FW: Irregularities of Bowman's membrane. *Am J Ophthalmol* 1966;62:227.
53. Wittehol-Post D, van Bijsterveld OP, Delleman JM: The honeycomb type of Reis-Bucklers' dystrophy of the cornea. Biometrics and interpretation. *Ophthalmologica* 1987;194:65.
54. Lohse E, Stock EL, Jones JC, et al: Reis-Bucklers' corneal dystrophy: Immunofluorescent and electron microscope studies. *Cornea* 1989;8:200.
55. Chan CC, Cogan DG, Bucci FS, Barsley D, Crawford MA: Anterior corneal dystrophy with dyscollagenosis. (Reis-Bucklers' type?). *Cornea* 1993;12:451.
56. Griffith DG, Fine BS: Light and electron microscopic observations in a superficial corneal dystrophy. *Am J Ophthalmol* 1967;63:1659.
57. Hoizan M, Wood L: Reis-Bücklers' corneal dystrophy. *Trans Ophthalmol Soc UK* 1971;91:41.
58. Kanai A, Kaufman HE, Polack FM: Electron microscopic study of Reis-Bücklers' dystrophy. *Ann Ophthalmol* 1973;5:953.
59. Perry HD, Fine BS, Caldwell CR: Reis-Bücklers' dystrophy. *Arch Ophthalmol* 1979;97:664.
60. Yamaguchi T, Polack F, Valenti J: Reis-Bücklers' corneal dystrophy. *Am J Ophthalmol* 1980;90:95.
61. Akiya S, Brown SI: The ultrastructure of Reis-Bucklers' dystrophy. *Am J Ophthalmol* 1971;72:549.
62. Yamaguchi T, Polack FM, Valenti L: Electron microscopic study of recurrent Reis-Bucklers' corneal dystrophy. *Am J Ophthalmol* 1980;90:95.
63. Rogers C, Cohen P, Lawless M: Phototherapeutic keratectomy for Reis-Bücklers' corneal dystrophy. *Aust NZ J Ophthalmol* 1993;21:247.
64. McDonnell PJ, Seiler T: Phototherapeutic keratectomy with excimer laser for Reis-Buckler's corneal dystrophy. *Refract Corneal Surg* 1992;8:306.
65. Lawless MA, Cohen PR, Rogers CM: Retreatment of undercorrected photorefractive keratectomy for myopia. *J Refract Corneal Surg* 1994;10(Suppl 2): 174.

66. Hahn TW, Sah WJ, Kim JH: Phototherapeutic keratectomy in nine eyes with superficial corneal diseases. *Refract Corneal Surg* 1994;9(Suppl 2):115–118.
67. Rapuano CJ, Laibson PR: Excimer laser phototherapeutic keratectomy for anterior corneal pathology. *CLAO J* 1994;20:253.
68. Orndahl M, Fagerholm P, Fitzsimmons T, Tengroth B: Treatment of corneal dystrophies with excimer laser. *Acta Ophthalmol* 1994;72:235.
69. Chamon W, Azar DT, Stark WJ, Reed C, Enger C: Phototherapeutic keratectomy. *Ophthalmol Clin North Am* 1993;6:399.
70. Wood TO, Fleming JC, Dotson RS, Cotten MS, et al: Treatment of Reis-Bücklers' corneal dystrophy by removal of subepithelial fibrous tissue. *Am J Ophthalmol* 1978;85:360.
71. Olson RJ, Kaufman HE: Recurrence of Reis-Bücklers' corneal dystrophy in a graft. *Am J Ophthalmol* 1978;85:349.
72. Caldwell DR: Postoperative recurrence of Reis-Bücklers' corneal dystrophy. *Am J Ophthalmol* 1978;85:567.
73. Grayson M, Wilbrandt H: Dystrophy of the anterior limiting membrane of the cornea (Reis-Bucklers type). *Am J Ophthalmol* 1966;63:345.
74. Kurome H, Noda S, Hayasaka S, Setogawa T: A Japanese family with Grayson-Wilbrandt variant of Reis-Bucklers' corneal dystrophy. *Jpn J Ophthalmol* 1993;37:143.
75. Fogle JA, Green WR, Kenyon KR: Anterior corneal dystrophy. *Am J Ophthalmol* 1974;77:529.
76. Thiel HJ, Behnke H: Eine Bisher Unbekannte Subepitheliale Herediteire Hornhautdystrophie. *Klin Monatsbl Augenheilkd* 1967;150:862.
77. Yamaguchi T, Polack FM, Rowsey JJ: Honeycomb-shaped corneal dystrophy: A variation of Reis-Bücklers' dystrophy. *Cornea* 1982;1:71.
78. Feder RS, Jay M, Yue BYJT, Stock EL, O'Grady RB, Roth SI: Subepithelial micinious corneal dystrophy: Clinical and pathological correlations. *Arch Ophthalmol* 1993;111:1106.
79. Rodrigues MM, Clahoun J, Harley RD: Corneal clouding with increased acid mucopolysaccharide accumulation in Bowman's membrane. *Am J Ophthalmol* 1975;79:916.
80. Streiff EB, Zwahlen P: Une famille avec bandelette de la cornee. *Ophthalmologica (Paris)* 1947;111:129.
81. Glees M: Ueber Familiaeres Auftreten der Primaeren, Bandfoermigen Hornhoatdegeneration. *Klin Monatsbl Augenheilkd* 1950;116:185.
82. Duke-Elder S, Leigh AG: *System of Ophthalmology*, Vol 8: *Diseases of the Outer Eye*. St Louis, Mosby–Year Book, 1965.
83. Akiya S, Brown S: Granular dystrophy of the cornea. *Arch Ophthalmol* 1970;8:179.
84. Schutz S: Hereditary corneal dystrophy. *Arch Ophthalmol* 1943;29:523.
85. Moller HU, Ridgway AEA: Granular corneal dystrophy Groenouw type 1: A report of a probable homozygous patient. *Acta Ophthalmol* 1990;68:97–101.
86. Andresen IL: Granular corneal dystrophy or Groenouw's disease type 1. A challenge to Norwegian biochemists, geneticists and ophthalmologists [in Norwegian]. *Tidskrift Norske Laegeforening* 1995;115:355.
87. Waardenburg PJ, Jonkers GH: A specific type of dominant progressive dystrophy of the cornea, developing after birth. *Acta Ophthalmol (Copenh)* 1961;39:919.
88. Forsius H, Erickson AW, Karna J, Tarkkanen A, Aurekoski H, et al: Granular corneal dystrophy with late manifestation. *Acta Ophthalmol (Copenh)* 1983;61:514.
89. Rodrigues MM, Gaster RN, Pratt MV: Unusual superficial confluent form of granular corneal dystrophy. *Ophthalmology* 1983;90:1507.
90. Haddad R, Font RL, Fine BS: Unusual superficial variant of granular dystrophy of the cornea. *Am J Ophthalmol* 1977;83:213.
91. Moller HU, Ehlers N, Bojsen-Moller M, Ridgway AE: Differential diagnosis between granular corneal dystrophy Groenouw type I and paraproteinemic crystalline keratopathy. *Acta Ophthalmol (Copenh)* 1993;71:552.
92. Garner A: Histochemistry of corneal granular dystrophy. *Br J Ophthalmol* 1969;53:799.
93. Iwamoto T, Stuart JC, Srinivasan BD, Mund ML, Farris RL, et al: Ultrastructural variations in granular dystrophy of the cornea. *Graefes Arch Clin Exp Ophthalmol* 1975;194:1.
94. Matsuo N, Fujiwara H, Ofuchi Y: Electron and light microscopic observations in a case of Groenouw's corneal dystrophy and gelatinous droplike dystrophy of the cornea. *Folia Ophthalmol Jpn* 1967;18:436.
95. Kuwabara Y, Akiya S, Obazawa H: Electron microscopic study of granular dystrophy, macular dystro-

phy, and gelatinous droplike dystrophy of the cornea. *Folia Ophthalmol Jpn* 1967;18:463.

96. Wittebol-Post D, van der Want JJ, van Bijsterveld OP: Granular dystrophy of the cornea (Groenouw type 1): Is the keratocyte the primary source after all? *Ophthalmologica* 1987;195:169.

97. Moller HU, Bojsen-Moller M, Schroder HD, Nelson ME, Vegge T: Immunoglobulins in granular corneal dystrophy Groenouw type 1. *Acta Ophthalmol* 1993;71:548.

98. Lyons CJ, McCartney AC, Kirkness CM, Ficker LA, Steele AD McG, Rice NSC: Granular corneal dystrophy. Visual results and pattern of recurrence after lamellar or penetrating keratoplasty. *Ophthalmology* 1994;101:1812.

99. Samson E: Granular dystrophy of the cornea: An electron microscopic study. *Am J Ophthalmol* 1965;59:1001.

104. Campos M, Nielsen S, Szerenyi K, et al: Clinical follow-up of phototherapeutic keratectomy for treatment of corneal opacities. *Am J Ophthalmol* 1993;115:433.

105. Stuart JC, Mund ML: Recurrent granular corneal dystrophy. *Am J Ophthalmol* 1975;79:18.

106. Tripathi R, Garner A: Corneal granular dystrophy: A light and electron microscope study of its recurrence in a graft. *Br J Ophthalmol* 1970;54:361.

107. Diaper CJ: Severe granular dystrophy: A pedigree with presumed homozygotes. *Eye* 1994;8(Pt 4):448.

108. Dark AJ, Thompson DS: Lattice dystrophy of the cornea. A clinical and microscopic study. *Br J Ophthalmol* 1960;44:257.

109. Reschmi CS, English FP: Unilateral lattice dystrophy of the cornea. *Med J Aust* 1971;1:966.

110. Raab MF, Blodi F, Boniuk M: Unilateral lattice dystrophy of the cornea. *Trans Am Acad Ophthalmol Otolaryngol* 1974;78:440.

111. Mehta RF: Unilateral lattice dystrophy of the cornea. *Br J Ophthalmol* 1980;64:53.

112. Ramsey RM: Familial corneal dystrophy lattice type. *Trans Am Ophthalmol* 1980;64:53.

113. Tsunoda I, Awano H, Kayama H, et al: Idiopathic AA amyloidosis manifested by autonomic neuropathy, vestibuloxoxhleopathy and lattice corneal dystrophy. *J Neurol Neurosurg Psychiatry* 1994;57:635.

114. Petrontos G, Kitsos G, Asproudies I, Melissamgos I, Psilas K: Association of progressive external ophthalmoplegia and lattice corneal dystrophy. *J Fr Ophthalmol* 1992;15:592.

115. Meretoja J: Familial systemic paramyloidosis with lattice dystrophy of the cornea, progressive crania neuropathy, skin changes and various internal symptoms: A previously unrecognized heritable syndrome. *Ann Clin Res* 1969;1:314.

116. Meretoja J: Comparative histopathological and clinical findings in eyes with lattice corneal dystrophy of two types. *Ophthalmologica* 1972;165:15.

117. Meretoja J: Genetic aspects of familial amyloidosis with corneal lattice dystrophy and cranial neuropathy. *Clin Genet* 1973;4:173.

118. Asaoka T, Amano S, Sunada Y, Sawa M: Lattice corneal dystrophy type II with familial amyloid polyneuropathy type IV. *Jpn J Ophthalmol* 1993;37:426.

119. Hida T, Tsobota K, Kigasawa K, et al: Clinical features of a newly recognized type of lattice corneal dystrophy. *Am J Ophthalmol* 1987;104:241.

120. Bowen RA, Hassard DTR, Wong VG, et al: Lattice dystrophy of the cornea as a variety of amyloidosis. *Am J Ophthalmol* 1970;7:822.

121. François J, Fehér J: Light microscopical and polarization optical study of lattice dystrophy of the cornea. *Ophthalmologica* 1972;164:1.

122. Hogan M, Alvarado S: Ultrastructure of lattice dystrophy of the cornea: A case report. *Am J Ophthalmol* 1967;64:656.

123. Smith M, Zimmerman L: Amyloid in corneal dystrophies. *Arch Ophthalmol* 1968;79:407.

124. Kivela T, Tarkkanen A, McLean I, Ghiso J, Frangione B, Haltia M: Immunohistochemical analysis of lattice corneal dystrophies types I and II. *Br J Ophthalmol* 1993;77:799.

125. Rodrigues MM, Rajgopalan S, Jones K, et al: Gelsoin immunoactivity in corneal amyloid, wound healing and macular and granular dystrophies. *Am J Ophthalmol* 1993;115:644.

126. Fogle JA, Kenyon KR, Syark WJ, Green WR: Defective epithelial adhesion in anterior corneal dystrophies. *Am J Ophthalmol* 1975;79:95.

127. Zechner EM, Croxatto JO, Mabran ES: Superficial involvement in lattice corneal dystrophy. *Ophthalmologica* 1986;193:19309.

128. Hida T, Proia AD, Kigasawa K, et al: Histopathologic and immunochemical features of lattice corneal dystrophy type III. *Am J Ophthalmol* 1987;104:249.

129. Mondino BJ, Raj CVS, Skinner M, et al: Protein AA and lattice corneal dystrophy. *Am J Ophthalmol* 1980;89:377.

130. Wheeler GE, Eiferman RA: Immunohistochemical identification of the AA protein in lattice dystrophy. *Exp Eye Res* 1983;36:181.

131. Bishop PN, Bonshek RE, Jones CJP, Ridgway AEA, Stoddart RW: Lectin binding sites in normal, scarred, and lattice dystrophy corneas.
132. Meisler DM, Fine M: Recurrence of the clinical signs of lattice corneal dystrophy (type 1) in corneal transplants. *Am J Ophthalmol* 1984;97:210.
133. Folberg R, Alfonso E, Croxatto JO, et al: Clinically atypical granular dystrophy with pathologic features of lattice-like amyloid deposits. A study of three families. *Ophthalmology* 1988;95:46.
134. Holand EJ, Daya SM, Stone EM, et al: Avellino corneal dystrophy: Clinical manifestations and natural history. *Ophthalmology* 1992;99:1564.
135. Yanoff M, Fine BS, Colosi NJ, Katowitz JA: Lattice corneal dystrophy. Report of an unusual case. *Arch Ophthalmol* 1977;95:651.
136. Stone EM, Mathers WD, Rosenwasser GOD, et al: Three autosomal dominant corneal dystrophies map to chromosome 5q. *Nature Genet* 1994;6:47.
137. Stock EL, Feder RS, O'Grady RB, Sugar J, Roth SI: Lattice corneal dystrophy type 111A: Clinical and histopathologic correlations. *Arch Ophthalmol* 1991;109:354.
138. Rosenwasser GOD, Sucheski BM, Rosa N, et al: Phenotypic variation in combined granular-lattice (Avellino) corneal dystrophy. *Arch Ophthalmol* 1993;111:1546.
139. Folberg R, Stone EM, Sheffield VC, Mathers WD: The relationship between granular, lattice type 1, and Avellino corneal dystrophies: A histopathologic study. *Arch Ophthalmol* 1994;112;1080.
140. Sassani J, Smith SG, Rabinowitz YS: Keratoconus and bilateral lattice-granular corneal dystrophies. *Cornea* 1992;11:343.
141. Cennamo G, Rosa N, Rosenwasser GOD, Sebastiani A: Phototherapeutic keratectomy in the treatment of Avellino dystrophy. *Ophthalmologica* 1994;208:198.
144. Klintworth GK, Smith CF: Macular corneal dystrophy: Studies of sulfated glycosaminoglycans in corneal explant and confluent stromal cell cultures. *Am J Pathol* 1977;89:167.
145. Hassell JR, Newsome DA, Krachmer JH, Rodrigues M: Macular corneal dystrophy: Failure to synthesize a mature keratan sulfate proteoglycan. *Proc Natl Acad Sci USA* 1980;77:3705.
146. Nakazawa K, Hassell JR, Hascall VC, Lohmander S, Newsome DA, Krachmer J: Defective processing of keratan sulfate in macular corneal dystrophy. *J Biol Chem* 1984;259:13751.
147. Bruner WE, Dejak TR, Grossniklaus HE, Stark WJ, Young E: Coneal alpha-galactosidase deficiency in macular corneal dystrophy. *Ophthalm Paediatr Genet* 1985;5:179.
148. Klintworth GK, Meyer R, Dennis R, et al: Macular corneal dystrophy—Lack of keratan sulfate in serum and cornea. *Ophthalmol Paediatr Genet* 1986;7:139.
149. Edward D, Thonar EJ-MA, Srinivasan M, et al: Macular corneal dystrophy of the cornea. A systemic disorder of keratan sulfate metabolism. *Ophthalmology* 1990;97:1194.
150. Quantock AJ, Meek KM, Ridgway AEA, Bron AJ, Thonar EJMA: Macular corneal dystrophy: Reduction in both corneal thickness and collagen interfibrillar spacing. *Curr Eye Res* 1990;9:393.
151. Santo RM, Yamaguchi T, Kanai A, Okisaka S, Nakajima A: Clinical and histopathologic features of corneal dystrophies in Japan. *Ophthalmology* 1995;102:557.
152. Yang CJ, SundarRaj N, Thonar EJ-MA, Klintworth GK: Immunohistochemical evidence of heterogeneity in macular corneal dystrophy. *Am J Ophthalmol* 1988;106:65.
153. Edward DP, Yue BYJT, Sugar J, et al: Heterogeneity in macular dystrophy. *Arch Ophthalmol* 1988;106:1579.
154. François J, et al: Ultrastructural findings in macular dystrophy (Groenouw type II). *Ophthalmic Res* 1975;7:80.
155. Jones ST, Zimmerman LE: Histopathologic differentiation of granular, macular, and lattice dystrophies of the cornea. *Am J Ophthalmol* 1961;51:394.
156. Garner A: Histochemistry of corneal macular dystrophy. *Invest Ophthalmol* 1969;9:473.
157. Francois J, Feher J: Light microscopical and polarization optical study of lattice dystrophy of the cornea. *Ophthalmologica* 1972;164:1.
158. Snip RC, Kenyon DR, Green RD: Macular corneal dystrophy: Ultrastructural pathology of the corneal endothelium and Desçemet's membrane. *Invest Ophthalmol* 1973;12:88.
159. Teng CC: Macular dystrophy of the cornea: A histochemical and electron microscopic study. *Am J Ophthalmol* 1966;62:436.
160. Livni N, Abraham FA, Zauberman H: Groenouw's macular dystrophy: Histochemistry and ultrastructure of the cornea. *Doc Ophthalmol* 1974;37:327.
161. Morgan G: Macular dystrophy of the cornea. *Br J Ophthalmol* 1966;50:57.
162. Klintworth GK, Vogel FS: Macular corneal dystrophy: An inherited acid mucopolysaccharide storage disease of corneal fibroblasts. *Am J Pathol* 1964;45:565.

163. Quantock AJ, Meek KM, Thonar EJMA, Assil KK: Synchrotron X-ray diffraction in atypical macular dystrophy. *Eye* 1993;7:779.
164. Yang CJ, SundarRaj N, Thonar EJ, Klintworth GK, et al: Immunohistochemical evidence of heterogeneity in macular corneal dystrophy. *Am J Ophthalmol* 1988;106:65.
165. Edward DP, Yue BY, Sugar J, Thonar EJ, et al: Heterogeneity in macular corneal dystrophy. *Arch Ophthalmol* 1988;106:1579.
166. Thonar EJ, Meyer RF, Dennis et al: Absence of normal keratan sulfate in the blood of patients with macular corneal dystrophy. *Am J Ophthalmol* 1986;102:561.
167. Klintworth GK, Smith CF: Abnormalities of proteoglycans and glycoproteins synthesized by corneal organ cultures derived from patients with macular corneal dystrophy. *Lab Invest* 1983;48:603.
168. Hassell JR: Defective conversion of a glycoprotein precursor to keratan sulfate proteoglycan in macular corneal dystrophy. In Hawkes S, Wang JL (eds): *Extracellular Matrix*. New York, Academic Press, 1982.
169. SundarRaj N, Barbacci-Tobin E, Howe WE, Robertson SM, Limetti G, et al: Macular corneal dystrophy: Immunohistochemical characterization using monoclonal antibodies. *Invest Ophthalmol Vis Sci* 1987;28:1678.
170. François J, Vicotria-Troncoso V, Maudgal PC, Victoria-Ihler A: Study of the lysosomes by vital stains in normal keratocytes and in keratocytes from macular dystrophy of the cornea. *Invest Ophthalmol Vis Sci* 1976;15:559.
171. Wu WCS, Stark WJ, Green WR: Corneal wound healing after 193-nm excimer laser keratectomy. *Arch Ophthalmol* 1991;109:1426.
172. Robin AL, Green WR, Lapsa TP, Hoover RE, Kelley JS: Recurrence of macular corneal dystrophy after lamellar keratoplasty. *Am J Ophthalmol* 1977;84:457.
173. Klintworth GK, Reed J, Stainer GA, Binder PS, et al: Recurrence of macular corneal dystrophy within grafts. *Am J Ophthalmol* 1983;95:60.
174. Delleman JW, Winkelman JE: Degeneratio corneae cristallinea hereditaria: A clinical, genetical, and histological study. *Ophthalmologia* 1968;155:409.
175. Bron AJ, Williams HP, Carruthers ME: Hereditary crystalline stromal dystrophy of Schnyder: Clinical features of a family with hyperlipoproteinemia. *Br J Ophthalmol* 1973;56:383.
176. Grop K: Clinical and histologic findings in crystalline corneal dystrophy. *Acta Ophthalmol (Copenh)* 1973;51(Suppl 120):52.
177. McCarthy M, Innis S, Dubord P, White V: Panstromal Schnyder corneal dystrophy. *Ophthalmology* 1994;101:895.
178. Williams HP, Bron AJ, Tripathi RC, Garner A: Hereditary crystalline corneal dystrophy with an associated blood lipid disorder. *Trans Ophthalmol Soc UK* 1971;91:31.
179. Garner A, Tripathi RC: Hereditary crystalline stromal dystrophy of Schnyder. II. Histopathology and ultrastructure. *Br J Ophthalmol* 1972;56:400.
180. Brownstein S, Jackson WB, Onerheim RM: Schnyder's crystalline corneal dystrophy in association with hyperlipoproteinemia; Histopathological and ultrastructural findings. *Can J Ophthalmol* 1991;26:273.
181. Ghosh M, McCulloch C: Crystalline dystrophy of the cornea: A light and electron microscopic study. *Can J Ophthalmol* 1977;12:321.
182. Rodrigues MM, Kruth HS, Krachmer JH, et al: Cholesterol localization in ultrathin frozen sections in Schnyder's corneal crystalline dystrophy. *Am J Ophthalmol* 1990;110:513.
183. Weller RO, Rodger FC: Crystalline stromal dystrophy: Histochemistry and ultrastructure of the cornea. *Br J Ophthalmol* 1986;64:46.
184. Rodrigues MM: Unesterified cholesterol in Schnyder's corneal crystalline dystrophy. *Am J Ophthalmol* 1987;104:147.
185. Lisch W, Weidle EG, Lisch C, Rice T, Beck E, Utermann G, et al: Schnyder's dystrophy: Progression and metabolism. *Ophthlamic Paediatr Genet* 1986;7:45.
186. Ehlers N, Mathiessen M: Hereditary crystalline dystrophy of Schnyder. *Acta Ophthalmol (Copenh)*: 1967;51:1.
187. Kaseras A, Price A: Central crystalline corneal dystrophy. *Br J Ophthalmol* 1970;54:659.
188. Fry WE, Pickett WE: Crystalline dystrophy of the cornea. *Trans Am Ophthalmol Soc* 1950;48:220.
189. Van Went JM, Wibaut F: Een Zeldzane erfelijke hoornvliesaandoening. *Ned Tijdschr Geenskd* 1924;68:2996.
190. Luxenburg M: Hereditary crystalline dystrophy of the cornea. *Am J Ophthlamol* 1967;63:507.
191. Weiss JS: Schnyder crystalline dystrophy sine crystals. Recommendation for a revision of nomenclature. *Ophthalmology* 1996;103:465.
192. Waring GO III, Rodrigues MM, Laibson PR:

Corneal dystrophies. I. Dystrophies of the epithelium, Bowman's layer and stroma. *Surv Ophthalmol* 1978;23:71.
193. Barchiesi BJ, Eckel RH, Ellis PP: The cornea and disorders of lipid metabolism. *Surv Ophthalmol* 1991;36:1.
194. Weller RO, Rodger FC: Crystalline stromal dystrophy: Histochemistry and ultrastructure of the cornea. *Br J Ophthalmol* 1980;64:46.
195. Freddo RF, Polack FM, Leibowitz HM: Ultrastructural changes in the posterior layers of the cornea in Schnyder's corneal crystalline dystrophy. *Cornea* 1989;8:170.
196. Burns RP, Connor W, Gipson I: Cholesterol turnover in hereditary crystalline corneal dystrophy of Schnyder. *Trans Am Ophthalmol Soc* 1978;76:184.
197. Mirshahi M, Mirshahi SS, Soria C, et al: Secretion of plasminogen activators and their inhibitors in corneal fibroblasts. Modification of this secretion in Schnyder's lens corneal dystrophy. *C R Acad Sci* 1990;311:253.
198. Sysi R: Xanthoma corneae as hereditary dystrophy. *Br J Ophthalmol* 1950;34:369.
199. Garner A, Tripathi RD: Hereditary crystalline stromal dystrophy of Schnyder. II. Histopathology and ultrastructure. *Br J Ophthalmol* 1972;56:400.
200. Nakaizumi K: A rare case of corneal dystrophy. *Nippon Ganka Gakkai Zasshi* 1914;18:949.
201. Akiya S, Ito I, Matsui M: Gelatinous drop-like dystrophy of the cornea. *Jpn J Clin Ophthalmol* 1972;56:815.
202. Nagataki S, Tanishima T, Sakomoto T: A case of primary gelatinous drop-like corneal dystrophy. *Jpn J Ophthalmol* 1972;16:107.
203. Weber FL, Babel J: Gelatinous drop-like dystrophy. *Arch Ophthalmol* 1980;98:144.
204. Gartry DS, Falcon MG, Cox RW: Primary gelatinous drop-like keratopathy. *Br J Ophthalmol* 1989;73:661.
205. Kirk HQ, Rabb M, Hattenhauer J, Smith R, et al: Primary familial amyloidosis of the cornea. *Trans Am Acad Ophthalmol Otolaryngol* 1973;77:411.
206. Stock EI, Kielar RA: Primary familial amyloidosis of the cornea. *Am J Ophthalmol* 1976;82:266.
207. Lewkojewa EF: Uber einen Fall primärer Degenerationamyloidose der Kornea. *Klin Monatsbl Augenheikd* 1930;85:117.
208. Ramsey MS, Fine BS: Localized corneal amyloidosis. *Am J Ophthalmol* 1972;75:560.
209. Santo RM, Yamaguchi T, Kanai A, Okisaka S, Nakajima A: Clinical and histopathologic features of corneal dystrophies in Japan. *Ophthalmology* 1995;102:557.
210. Kanai A, Kaufman HE: Electron microscopic studies of primary band-shaped keratopathy and gelatinous drop-like corneal dystrophy in two brothers. *Ann Ophthalmol* 1982;14:535.
211. Matsui M, Ito K, Akiua S: Histochemical and electron microscopic examinations on so-called gelatinous drop-like dystrophy of the cornea. *Folia Ophthalmol Jpn* 1972;23:466.
212. Mondino BJ, Rabb MF, Sugar J, et al: Primary familial amyloidosis of the cornea. *Am J Ophthalmol* 1981;92:732.
213. Sakuma A, Yokoyama T, Katou K, Kanai A: Lamellar keratoplasty combined with keratoepithelioplasty in four cases of recurrent gelatinous drop-like corneal dystrophy [in Japanese]. *Rinsho Ganka* 1991;45:527.

CHAPTER 7
Anterior Corneal Dystrophies: PTK Techniques

Sandeep Jain, Dimitri T. Azar, Walter J. Stark, M.D.

INTRODUCTION

The excimer laser ablates corneal tissue with submicron precision without significant injury to non-ablated tissue.[1-3] It may be used to remove the anterior layers of the cornea to treat corneal opacities and to smooth corneal surface irregularities. The depth and shape of excimer laser ablative photodecomposition can be accurately controlled,[3-6] allowing for exact removal of the stroma and providing a relatively smooth base for reepithelialization.[1,3,5,6,8-15]

Surface irregularities and opacities resulting from dystrophies and degenerations limited to the anterior corneal layers can be improved with phototherapeutic keratectomy (PTK), especially if associated with significant visual impairment due to scarring and recalcitrant, recurrent corneal erosions.[5]

Conventional surgical methods of treating scarring due to corneal dystrophies include lamellar and penetrating keratoplasty. The disadvantages of such procedures include long rehabilitation time, unpredictable refractive results, and recurrence of pathology in the graft. Patients with superficial corneal lesions, as in epithelial and basement membrane dystrophies (Schnyder, Reis-Bückler, lattice, and Meesmann), respond well to PTK, obviating the need for conventional invasive surgery.

Conventional surgical methods of treating recurrent corneal erosions include manual epithelial débridement and anterior stromal puncture. Patients suffering from recalcitrant, recurrent corneal erosions (not relieved by conventional surgery) may benefit from excimer PTK (see Chapter 9). The treatment depth is relatively minimal and is usually limited to Bowman's layer. Accordingly, a significant postoperative hyperopic shift is not observed and corneal wound healing is less prolonged.

SURGICAL TECHNIQUE FOR TREATING CORNEAL DYSTROPHIES

Preoperative Preparation

Preoperative evaluation includes visual acuity, visual potential (evaluated with the pinhole, hard contact lens, and potential acuity meter), pupil size, slit lamp biomicroscopy, and dilated fundus examination. The laser is calibrated before each treatment to ensure optimal performance. The overall operation of the laser is confirmed by ablating a standard treatment into a caliberation plate made of a polymethyl methacrylate (PMMA) test block or other material, depending on the laser used. The appropriate corneal ablation rate is determined using nomograms and entered into the laser computer program.

Currently, most procedures are performed under topical anesthesia. Before treatment, the plane of the corneal surface is determined by focusing the microscope at high magnification. Poor centration of laser treatment may result in a suboptimal outcome.

Epithelial Removal (Laser Ablation vs. Manual Debridement)

The decision on whether to ablate using the laser or to remove the epithelium manually is based on the smoothness of the epithelium relative to the envisioned smoothness of the anterior stromal surface. If the anterior stromal surface is judged to be irregular, the epithelium is ablated with the laser. If the anterior stromal surface is judged to be smooth, the epithelium may be removed manually with a Bard-Parker blade.[5]

Laser Epithelial Ablation

Two methods of laser epithelial ablation have been described, differing in the use of masking fluids over the epithelium and determination of the depth of ablation.

The thickness of the corneal opacity, including the overlying epithelium, is measured with an optical pachymeter. PTK is then performed, set at the measured depth, through the intact epithelium, without the use of masking fluids.[7] Because the anterior epithelial surface is usually smooth in eyes with corneal dystrophies, while Bowman's membrane is nodular and irregular, ablating through the intact epithelium is recommended as part of the procedure, rather than first removing the epithelium. In this fashion, the epithelium is used as a masking agent allowing a smooth postoperative ablated surface.

Other investigators[8,9] have used a different technique to ablate the epithelium with laser. Masking fluid (0.5% hydroxymethylcellulose) is placed on the cornea, and the fluid and epithelium ablated until the blue fluorescence disappears. Disappearance of the blue fluorescence marks the point at which all of the epithelium has been ablated and stromal ablation begins. It is unclear, however, whether the use of masking fluid in this fashion creates a smoother postoperative surface.

Manual Epithelial Debridement

Most surgeons prefer to remove the epithelium manually in treating recurrent corneal erosions. However, for corneal dystrophies like Reis-Bücklers' dystrophy, manual epithelial débridement is not preferred. Sher et al. removed the epithelium manually prior to PTK in a patient with Reis-Bücklers' dystrophy.[10] Ornndahl et al. removed the epithelium manually in 14 of 33 patients with corneal dystrophies.[11] Difficulties in manual removal in dystrophies such as in Reis-Bücklers were noted by Lawless et al. They found the epithelium in patients with Reis-Bücklers' dystrophy to be unstable and to come off in a patchy manner when removed mechanically;[12] further, it was difficult to determine when removal was complete. Therefore, they preferred to remove the epithelium with the laser.

If the epithelium is removed manually prior to stromal ablation, a masking fluid (1% hydroxymethylcellulose, 0.5% Tetracaine or Tears Naturale II) is usually applied to improve the smoothness of the postoperative ablated surface. The fluid in valleys prevents ablation of underlying tissue, leaving the exposed peaks to be ablated. A highly viscous fluid (2% hydroxymethylcellulose or Healon) does not cover an irregular surface uniformly and tends to partially cover peaks as well as valleys. A fluid with low viscosity tends to expose both peaks and valleys. In addition to creating a smooth corneal surface, masking agents may reduce the amount of induced hyperopia (see Chapter 8).

Stromal Ablation

The depth of stromal ablation is determined preoperatively and corresponds to the depth of the stromal scar, as measured with the pachymeter. Some surgeons have performed slit lamp examinations during ablation to evaluate the result of treatment and plan additional treatments. For patients with granular dystrophy, the aim is to ablate most of the areas of diffuse haze between the granular hyaline deposits, not necessarily all the granular deposits (Figures 7.1 and 7.2).

The treatment depth in patients with recalcitrant, recurrent corneal erosions is relatively minimal (5–10) μm) and is usually limited to Bowman's layer. Since the pathology in corneal dystrophies is diffuse, a large (5.5–6.5 mm) beam diameter is used for stromal ablation.

A transition zone is usually created during stromal ablation. It is intended to allow smooth, uniform reepithelialization over the ablation bed. This procedure is referred to as *standard taper* ablation. Sher et al. used a *smoothing* technique in their early cases, wherein the eye was moved in a circular manner under the laser beam.[10] A similar *polish* technique was in the clinical trials with the Summit excimer laser. The surgeon moved the patient's head in a brisk, controlled circular manner under the laser beam to polish the corneal surface.[13,14] Stark et al. later described a *modified taper* technique, wherein the surgeon attempts to decrease central flattening by moving the eye under the laser

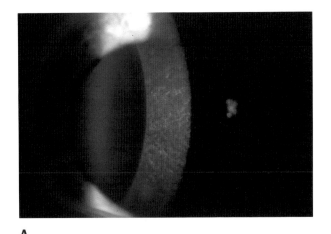

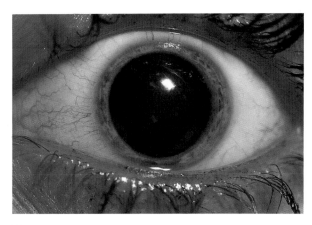

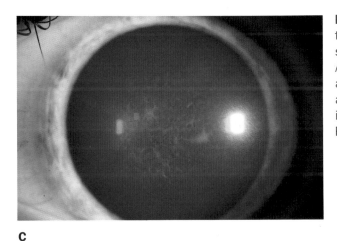

Figure 7.1. A 26-year-old woman with Reis-Bücklers' dystrophy presenting with 20/40 vision and a 2+ mean haze score. A: Clinical appearance of the cornea prior to PTK. B: Appearance 1 month following surgery, with 20/25 visual acuity. C: Twelve months after surgery, with similar visual acuity. Note that the recurrence of Reis-Bücklers' dystrophy is difficult to differentiate from scarring following excimer PTK.

in a circular fashion, treating the circumference of the ablation zone with a 200-μm-deep, 2-mm-diameter spot size.[1,15,16] This edge modification creates a ring-shaped ablation pattern at the periphery of the field to reduce the hyperopic shift seen after PTK. When corneal opacities or irregularities are associated with myopic refractive errors, a combination of PTK and photorefractive keratectomy (PRK) should be considered. By allowing for approximately 1 D of hyperopic shift for every 20 μm of stromal ablation, the need for PRK to correct any myopic refractive error is reduced. Talamo et al. have described clinical treatment strategies for optimizing the successful application of PTK to the treatment of superficial corneal pathology.[17] (See Chapters 8 and 10.)

Postoperative Management

Postoperative medications include prophylactic antibiotics and anti-inflammatory agents. Sub-Tenon's injection of gentamicin and dexamethasone is given immediately after surgery. After topical application of antibiotic ointment (bacitracin and erythromycin) and instillation of a cycloplegic agent (homatropine), the eye is patched. Alternatively, a therapeutic soft contact lens is applied, with frequent application of comparable medications. Since the patient may experience severe pain in the first 24 h, topical nonsteroidal anti-inflammatory agents and systemic sedative-analgesics may be needed for the first few days. The use of topical nonsteroidal agents like 0.1% diclofenac sodium (Voltaren

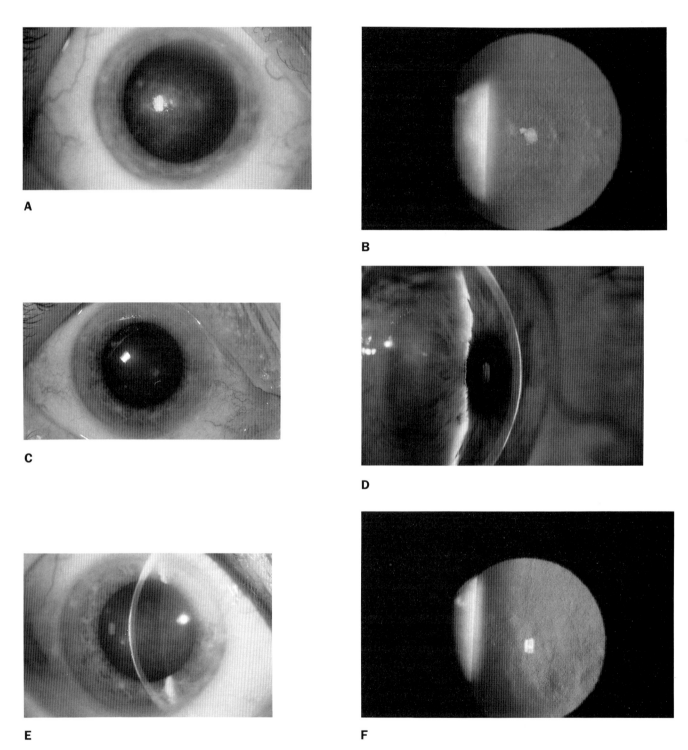

Figure 7.2. A 57-year-old man with Reis-Bücklers' dystrophy resulting in 20/80 vision who underwent PTK. *A, B:* Preoperative appearance by direct illumination and by retroillumination, respectively. *C:* Postoperative appearance of the eye 1 month postoperatively; Visual acuity improved to 20/50. *D:* Six months postoperatively, visual acuity regressed to 20/70. *E, F:* Appearance of the cornea by direct illumination and retroillumination, respectively, 20 months postoperatively, with similar visual acuity.

Ophthalmic, CIBA Vision Ophthalmics, Atlanta, GA) in the immediate postoperative period significantly reduces postoperative pain. A steroid regimen is instituted, comprising of an initial intensive therapy with 0.1% fluoromethanolone and subsequent titration of the dosage and duration of therapy with the postoperative effect. In patients with an uneventful postoperative course, steroids are tapered over a period 1–5 months. Steroid treatment is restarted or the dosage increased if clinically significant corneal haze occurs. The benefits of continued steroid drops may be outweighed by the potential side effects. Potential complications of long-term steroid treatment include increased susceptibility to infections, increased intraocular pressure, reactivation of herpes simplex, cataract formation, and reduced wound strength.

Patients are examined every 24 to 48 h until reepithelialization occurs and then at 1 month, 3 months, 6 months, 12 months, and 24 months. In most patients, reepithelialization occurs within 1 week. Following reepithelialization, the postoperative examination at each visit includes symptomatic evaluation, a detailed anterior segment examination, and slit lamp biomicroscopy. In addition, all measurements (visual acuity, ocular pressure, etc.) taken during the preoperative evaluation are repeated.

At each postoperative visit, corneal haze is graded subjectively using slit lamp biomicroscopic examination:[1] 0 = clear; 0.5 = barely detectable; 1.0 = mild, not affecting refraction; 1.5 = mildly affecting refraction; 2.0 = moderate, refraction possible but difficult; 3.0 = opacity preventing refraction, anterior chamber easily viewed; 4.0 = impaired view of the anterior chamber; and 5.0 = inability to see the anterior chamber. The cornea is divided into five hypothetical layers (superficial and deep epithelium, anterior and posterior stroma, and endothelium), and each layer is graded separately.

CLINICAL OUTCOME AFTER PTK FOR CORNEAL DYTROPHIES

Background

We have reviewed the clinical outcomes of approximately 600 eyes, treated for various corneal diseases with excimer laser PTK. They have been reported in 26 published articles and compiled in a previous publication.[5] (Table 7.1). Four excimer lasers have been used: Twenty/Twenty (VISX, Inc., Santa Clara, CA), LV 2000 (Taunton Technologies Co., Monroe, CT [now VISX 20/15]), ExciMed 200 (Summit Technologies, Inc., Waltham, MA), and MEL 50 (Aesculap Meditec, Heroldsberg, Germany).[5] Thirty-six percent (203) of the eyes were treated for recurrent corneal erosions, 20% (110) for corneal dystrophies, 16% (89) for corneal scars, 28% (157) for corneal irregularities, and 1 eye for recurrent corneal intraepithelial dysplasia. Success rates refer to the functional improvement of several parameters (visual acuity, corneal clarity, patient comfort, or cosmetic appearance) based on the goals of treatment. Due to the heterogeneity of the corneal pathologies treated with PTK, no single parameter can reliably estimate the success rate.

Functional improvement was achieved in 77% of the 203 eyes after the initial treatment of recurrent epithelial erosions with PTK. In two studies the success rate improved to 95% (107/113) with retreatment.

Corneal dystrophies and degenerations that have been treated with PTK to improve visual function or comfort include dystrophies of the epithelium and basement membrane (map-dot-fingerprint and Meesmann's [Table 7.2]), dystrophies of Bowman's layer (Reis-Bücklers' [Table 7.3]), stromal dystrophies (granular [Table 7.4], lattice [Table 7.5], gelatinous drop-like, macular, Schnyder), endothelial dystrophies (Fuch's), and Salzmann's nodular degeneration. Greater success has been achieved in treating superficial corneal dystrophies like map-dot-fingerprint (100%), Meesmann's (100%), and Reis-Bücklers' dystrophies (100%) than in treating stromal dystrophies like granular (67%), lattice (92%), and Schnyder's dystrophies (67%). Patients with recurrent granular or lattice dystrophy in a graft have relatively superficial lesions (Figures 7.2 to 7.5). The success rate in these cases is very high and is similar to that for primary Reis-Bücklers' dystrophy, in which the deposits are limited to Bowman's layer (Figures 7.6 and 7.7). Macular dystrophy and calcific scars carry especially high complication and Lower success rates (Figures 7.8 to 7.10).

Table 7.1. PTK in Corneal Diseases

Pathology	Excimer Laser Used				No. of Eyes	Follow-up (months)	Success (%)
	VisX	Summit	Meditec	Taunton			
Recurrent epithelial erosions	+	+	+	+	203	16.5	77%
Corneal dystrophies							
Reis-Bücklers'	+	+	−	+	32	6.2	100%
Lattice	+	+	−	−	25	11	92%
Granular	+	+	−	+	20	7.2	70%
Salzmann's nodular	+	+	−	+	12	8.6	67%
Map-dot-fingerprint	+	+	−	−	10	5.2	100%
Schnyder's	+	+	−	−	6	6	67%
Avellino	−	−	+	−	1	12	100%
Gelatinous drop-like	−	+	−	−	1	4	100%
Meesmann's	+	+	−	−	2	9	100%
Fuch's endothelial	+	+	−	−	1	9	100%
Corneal scars							
Postinfectious	+	+	−	+	12	8.5	50%
Posttraumatic	+	+	−	+	23	11.1	61%
Herpetic	+	+	−	+	24	15	71%
Trachomatous	−	+	−	−	3	6	67%
Pterygium	+	+	−	+	10	11.2	80%
Stevens-Johnson syndrome	−	+	−	−	3	11.7	67%
Contact lens wear related	+	+	−	−	2	3	50%
Unknown etiology	+	+	−	−	12	14.6	50%
Corneal irregularities							
Band keratopathy	+	+	−	+	136	12.3	91%
Apical scars in keratoconus	+	+	−	−	21	9.7	81%
Corneal intraepithelial dysplasia	−	−	+	−	1	26	100%

Modified with permission from: Azar DT, Jain S, Stark W.: Phototherapeutic Keratectomy. In *Refractive Surgery*. Azar D.T. (ed.) Appleton & Lange, 1996, pp. 508.

Table 7.2. Recurrent Corneal Erosions

Study	Laser	No. of Eyes	Phototherapeutic Keratectomy	Follow-up (months)	Success After Treatment	
					Initial	Retreatment
Ohman[18]	Summit VISX	76	Epithelial débridement and 3- or 5-μm PTK 20-μm PTK through intact epithelium	16.3	56 (74%)	70 (92%)
Dausch[19]	Meditec (800 mJ/cm^2)	74	1- to 3-μm PTK at epithelial defect sites 30- to 40-μm PTK at marginal epithelium	21.1	55 (74%)	—
Fagerholm[20]	Summit	37	Epithelial débridement and 3-μm PTK	11.8	31 (84%)	37 (100%)
Foster[21]	Summit	9	Epithelial débridement and 3- to 4-μm PTK	6	8 (89%)	—
Rapuano[22,23]	VISX	3	—	9	3 (100%)	—
John[24]	Summit	2	Epithelial débridement and 3- to 4-μm PTK	18	2 (100%)	—
Sher[10]	Taunton	1	Epithelial débridement and 30-μm PTK (simultaneous corneal scar removal)	10	1 (100%)	—
Hersh[9]	Summit	1	Epithelial débridement and 3.8-μm PTK	4	1 (100%)	—

Reproduced with permission from: Azar DT, Jain S, Stark W: Phototherapeutic Keratectomy. In *Refractive Surgery*. Azar DT (ed.). Appleton and Lange, 1996, pp. 505.

Table 7.3. Reis-Bücklers' Dystrophy

Study	Laser	No. of Eyes	Phototherapeutic Keratectomy	Follow-up (months)	Success (%)	Hyperopic Shift Eyes (%)	Hyperopic Shift Range (diopters)
Rogers[8]	Summit	11	Hydroxymethylcellulose (0.5%) Fluorescence-guided epithelial ablation PTK (19–63 μm); focal technique	6	100	100	0.25–7.0
Lawless[12]	Summit	9	Hydroxymethylcellulose (0.5%) Fluorescence-guided epithelial ablation PTK 18–60 μm); focal technique	6	100	78	0.25–8.0
Hahn[25]	Summit	2	Hydroxymethylcellulose (1.0%) Mechanical epithelial débridement PTK (50 μm); focal technique	10	100	100	0.5–1.5
Hersh[9]	Summit	2	Hydroxymethylcellulose (1.0%) Epithelial débridement/ablation PTK smoothing technique	4	100	100	Up to 7.0
McDonnell[7]	Summit	1	Epithelial ablation PTK (100 μm); smoothing technique PRK −4.0 D	6	100	100	10.75
	VISX	1	Epithelial ablation PTK (100 μm); focal technique	6	100	100	3.8
Sher[10]	Taunton	1	Mechanical epithelial débridement PTK (50 μm) (Combined myopic and hyperopic cut)	6	100	100	7.25
Rapuano[22,23]	VISX	1	Mechanical epithelial débridement PTK focal technique	3	100	100	1.25
Orndahl[11]	Summit VISX	2	—	9	100	—	—
Stark[1]	VISX	2	—	—	100	—	—

Reproduced with permission from: Azar DT, Jain S, Stark W: Phototherapeutic Keratectomy. In *Refractive Surgery*. Azar DT (ed.). Appleton and Lange, 1996, pp. 510.

Table 7.4. Granular Dystrophy

Study	Laser	No. of Eyes	Phototherapeutic Keratectomy	Follow-up (months)	Success (%)	Hyperopic Shift Eyes (%)	Hyperopic Shift Range (diopters)
Hahn[25]	Summit	2	Hydroxymethylcellulose (1.0%) Mechanical epithelial débridement PTK (40 μm); focal technique	8.3	66	66	Up to 2.0
Sher[10]	Taunton	2	Mechanical epithelial débridement PTK (50 μm) (Combined myopic and hyperopic cut)	6	0	100	0.3–1.1
Rapuano[22,23]	VISX	6	Epithelial débridement/ablation PTK disciform/elliptical	8.3	83	63	0.62–2.0
Orndahl[11]	Summit VISX	4	Fluorescence-guided epithelial ablation PTK (45 μm)	12	75	—	Up to 2.0
Stark[1]	VISX	4	PTK standard taper/modified taper	—	75	—	—
Campos[26]	VISX	1	Epithelial débridement PTK (110 μm); disciform	24	100	—	—

Table 7.5. Lattice Dystrophy

Study	Laser	No. of Eyes	Phototherapeutic Keratectomy	Follow-up (months)	Success (%)	Hyperopic Shift Eyes (%)	Hyperopic Shift Range (diopters)
Hersh[9]	Summit	1	Hydroxymethylcellulose (1.0%) Epithelial débridement/ablation PTK smoothing technique	4	100	0	—
Orndahl[11]	Summit VISX	11	Fluorescence-guided epithelial ablation PTK (45 μm)	12	90	—	Up to 2.0
Stark[1]	VISX	11	PTK standard/modified taper	—	90	—	—
Campos[26]	VISX	2	Mechanical epithelial débridement PTK (100–110 μm); disciform	10	100	100	3.0–8.2

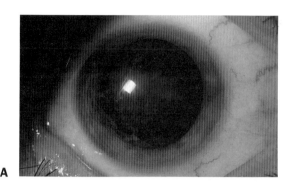
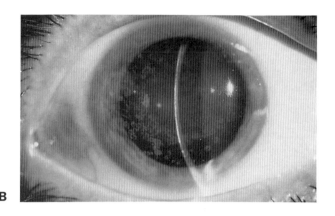

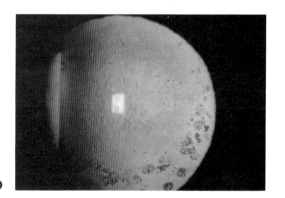
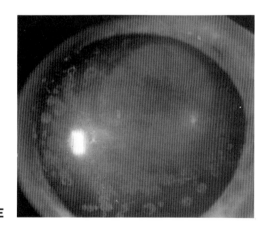
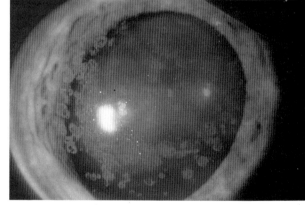
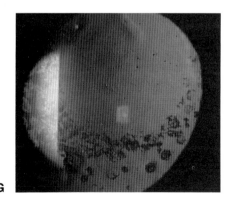

Figure 7.3. A 42-year-old man with granular dystrophy who underwent two PTK operations. *A:* Preoperative appearance with 20/200 visual Acuity. *B, C:* Clinical appearance by direct illumination and retroillumination, respectively, 3 months after the first PTK. Visual acuity became 20/1000. *D:* Six months postoperatively, with visual acuity of 20/600. *E:* Twelve months postoperatively, with visual acuity of 20/200. *F, G:* Direct illumination and retroillumination of the eye, respectively, 12 months after the second PTK.

Anterior Corneal Dystrophies: PTK Techniques

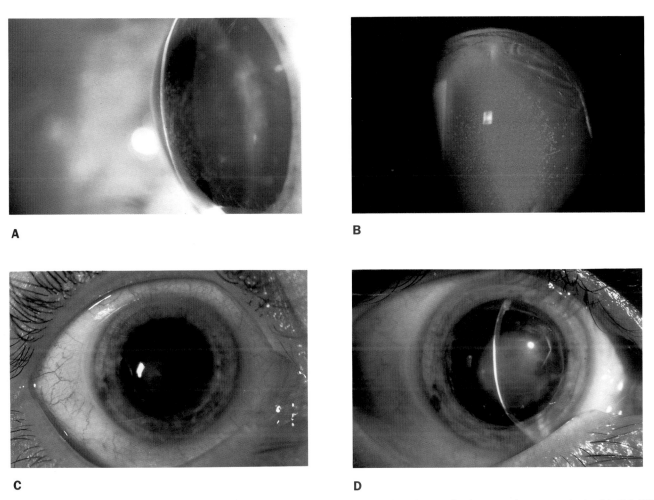

Figure 7.4. A 48-year-old woman with a history of recurrent granular dystrophy in a graft presenting with 20/200 vision and a 1+ mean haze score. *A:* Clinical appearance of the cornea prior to PTK by direct illumination. *B:* Clinical appearance of the cornea by retroillumination. *C:* Appearance 3 months after PTK. Visual acuity improved to 20/80, with no corneal haze. *D:* One year after PTK; note the increased scarring in the area of treatment.

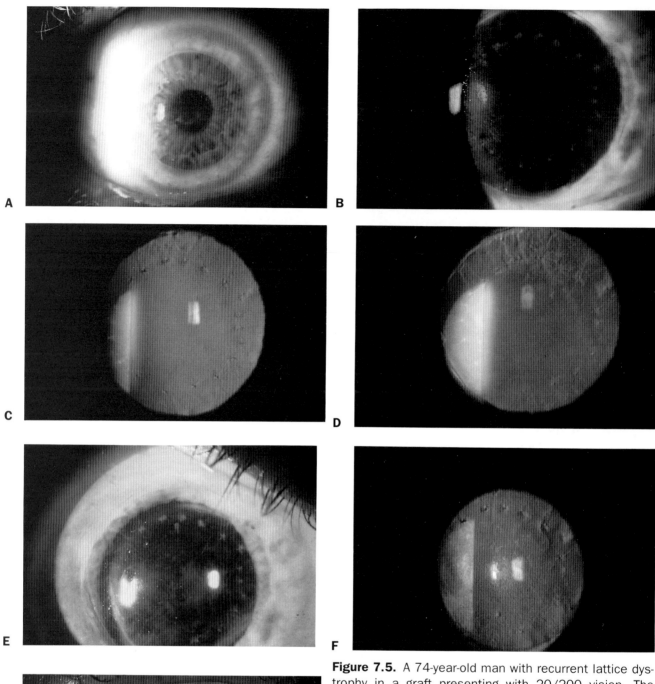
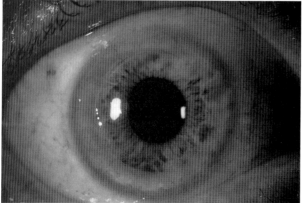

Figure 7.5. A 74-year-old man with recurrent lattice dystrophy in a graft presenting with 20/200 vision. The patient underwent PTK and a second treatment 4.5 years after the first treatment due to recurrence. The patient is the father of the patients discussed in Figure 4.6. *A:* Preoperative appearance. *B, C:* At 5 months after surgery by direct illumination and retroillumination, respectively. Visual acuity improved to 20/70. *D:* Retroillumination of the eye 18 months postoperatively, with similar visual acuity. *E, F:* Photographs of the eye by direct illumination and retroillumination, respectively, prior to the second PTK. The patient presented with 20/160 vision. *G:* Six months after the second PTK with visual acuity of 20/100 was achieved.

Anterior Corneal Dystrophies: PTK Techniques

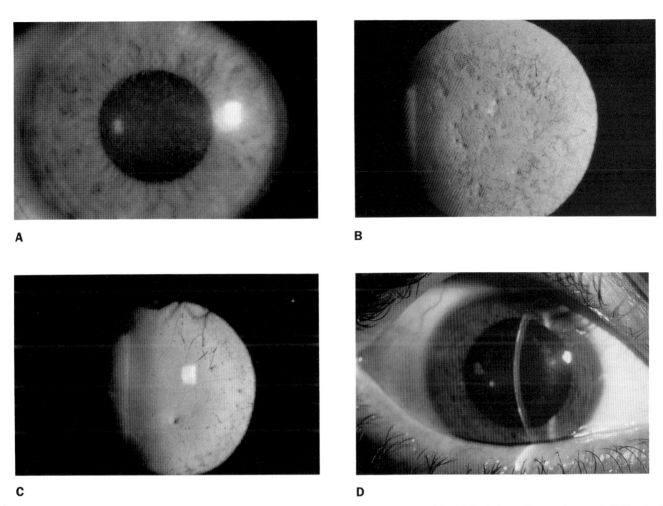

Figure 7.6. A 27-year-old woman with lattice dystrophy presenting with 20/200 vision who underwent PTK. *A, B:* Preoperative appearance of the lesion by direct illumination and retroillumination, respectively. *C:* Retroillumination of the eye after 1 month surgery, with visual acuity of 20/80. *D:* A slit lamp photograph of the eye at 18 months, with similar visual acuity.

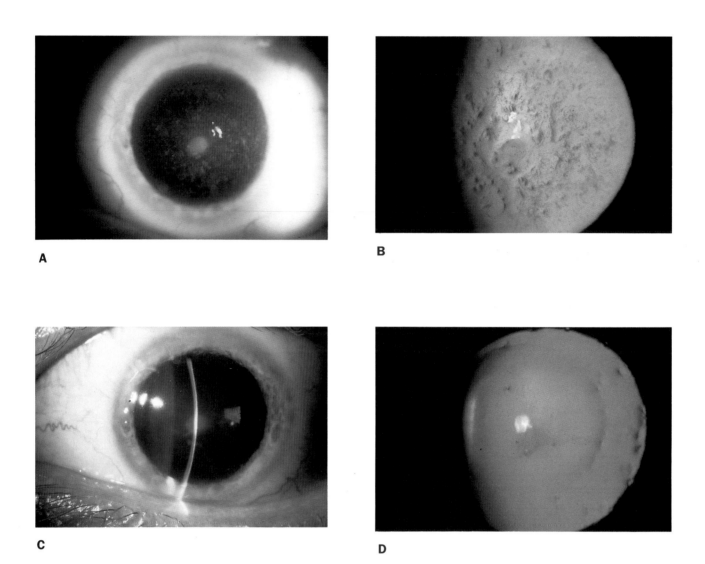

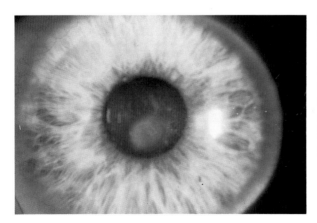

Figure 7.7. A 34-year-old man with lattice dystrophy resulting in 20/80 vision who underwent PTK. *A, B:* Preoperative appearance by direct illumination and retroillumination, respectively. *C, D:* Clinical appearance of the cornea by direct illumination and retroillumination, respectively, 1 month after surgery, with visual acuity of 20/50. Note the reduction in amyloid deposits in the area of treatment. *E:* Six months following PTK. Visual acuity improved to 20/40 in spite of the presence of diffuse haze.

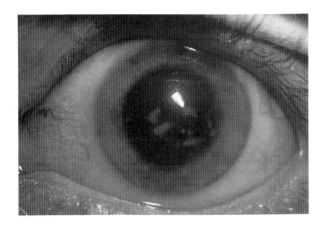
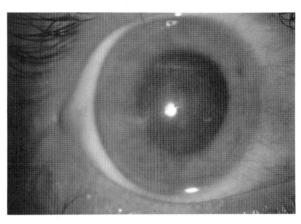

Figure 7.8. A 32-year-old woman with macular dystrophy resulting in 20/50 visual acuity. *A:* Clinical appearance prior to PTK. *B:* Appearance 3 months after surgery; visual acuity become 20/200.

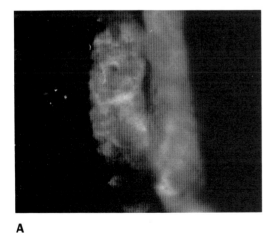
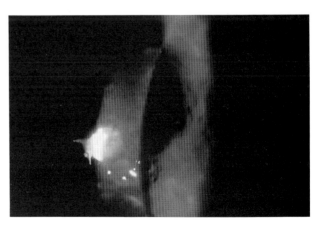

Figure 7.9. A 19-year-old man with calcific corneal scars who suffered from a marked decrease in vision, with 20/3000 visual acuity and a 3+ mean haze score. The patient underwent PTK. *A:* Preoperative photograph of the eye showing the lesion. *B:* Photograph of the eye 6 weeks after PTK. Although the mean haze score was reduced to zero 1 year after surgery, visual acuity failed to improve. The benefit of PTK in treating calcific corneal scars is questionable.

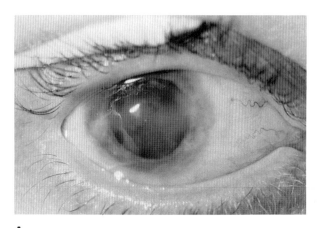
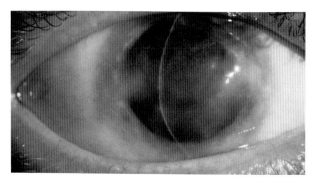
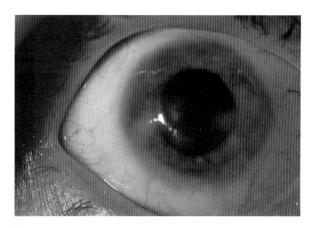

Figure 7.10. A 28-year-old woman with epithelial scars. *A:* Preoperative appearance of the eye, with 20/600 visual acuity and a 2+ mean haze score. *B:* Clinical appearance 1 month after PTK, with minimal improvement. *C:* Seven months after surgery, with improvement in visual acuity to 20/300 and in the mean haze score to 1+.

SIDE EFFECTS AND COMPLICATIONS

Multiple potential side effects of PTK for corneal dystrophies have been reported.[1,2,27–53] These will be discussed in detail in Chapter 10. Delayed corneal wound healing can follow PTK.[1,2] Stark et al. reported on two patients in whom corneal reepithelialization took 3–4 weeks compared to 1 week or less for other patients.[1] Delayed wound healing following PTK may be associated with corneal haze, recurrent erosions, infections, corneal ulcers, and persistent epithelial defects.[1,27–41]

Postoperative pain may be severe during the first 24 to 48 h. In some patients, moderate to severe pain may occur for several days after surgery, and it is generally relieved by the time the epithelium heals.

Flattening of the central cornea appears to be the principal side effect of PTK. Hyperopia is induced as a result of the corneal flattening. Its correction may require the use of a contact lens postoperatively. The use of appropriate masking agents may minimize the hyperopic shift. Other strategies include Stark and Azar's modified taper technique[1,15,16,41] or Sher's technique of preprogramming the Taunton laser system to cut a secondary hyperopic correction (*combined* ablation).[10]

If the eyes do not improve significantly after PTK, the patient may need to undergo more invasive treatment such as corneal transplantation.[1,5] Many patients who are currently treated with PTK to reduce the chance of needing penetrating keratoplasty (PK) end up requiring PK. With further advances in our technique and refinement of the indications for PTK, the need for PK after PTK will be minimized.

REFERENCES

1. Stark WJ, Chamon W, Kamp MT, Enger CL, Rencs EV, Gottsch JD: Clinical follow-up of 193-nm ArF excimer laser photokeratectomy. *Ophthalmology* 1992;99:805–811.
2. Salz JJ, Maguen E, Macy JI, Papaioannou T, Hofbauer J, Nesburn AB: One-year results of excimer laser photorefractive keratectomy for myopia. *Refract Corneal Surg* 1992;8:270–273.
3. Gaster RN, Binder PS, Coalwell K, Berns M, McCord RC, Burstein NL: Corneal surface ablation by 193 nm excimer laser and wound healing in rabbits. *Invest Ophthalmol Vis Sci* 1989;30:90–97.
4. Trokel SL, Srinivasan R, Braren B: Excimer laser surgery of the cornea. *Am J Ophthalmol* 1983;96:710–715.
5. Azar DT, Jain S, Stark WJ. Phototherapeutic Keratectomy. In *Refractive Surgery*. Azar DT (Ed) Appleton & Lange, 1997, pp. 501–517.
6. Marshall J, Trokel S, Rothery S, Krueger RR: Photoablative reprofiling of the cornea using an excimer laser: Photorefractive keratectomy. *Lasers Ophthalmol* 1986;1:23–44.
7. McDonnell PJ, Seiler T: Phototherapeutic keratectomy with excimer laser for Reis-Bücklers' corneal dystrophy. *Refract Corneal Surg* 1992;8:306–310.
8. Rogers C, Cohen P, Lawless M: Phototherapeutic keratectomy for Reis Bücklers' corneal dystrophy. *Aust NZ J Ophthalmol* 1993;21:247–250.
9. Hersh PS, Spinak A, Garrana R, Mayers M: Phototherapeutic keratectomy: Strategies and results in 12 eyes. *Refract Corneal Surg* 1993;(Suppl 2):90–95.
10. Sher NA, Bowers RA, Zabel RW, et al: Clinical use of 193-nm excimer laser in the treatment of corneal scars. *Arch Ophthalmol* 1991;109:491.
11. Orndahl M, Fagerholm P, Fitzsimmons T, Tengroth B: Treatment of corneal dystrophies with excimer laser. *Acta Ophthalmol* 1994;72:235–240.
12. Lawless MA, Cohen PR, Rogers CM: Retreatment of undercorrected photorefractive keratectomy for myopia. *J Refract Corneal Surg* 1994;10(Suppl 2):174–177.
13. Thompson V, Durrie DS, Cavanaugh TB: Philosophy and technique for excimer laser phototherapeutic keratectomy [review]. *Refract Corneal Surg* 1993;9(Suppl 2):81–85.
14. Durrie DS, Schumer JD, Cavanaugh T: Phototherapeutic keratectomy: The VISX experince. In Salz JJ, McDonnell PJ, McDonald MB (eds): *Corneal Laser Surgery*. St Louis, Mosby–Year Book, 1995, pp 227–235.
15. Azar DT, Chamon W, Stark WJ: Phototherapeutic keratectomy. In Stenson S (ed): *Surgical Management in External Diseases of the Eye*. Tokyo, Igaku-Shoin, 1996, pp. 00–00.
16. Azar DT, Jain S, Woods K, et al: Phototherapeutic keratectomy: The VISX Experience. In Salz JJ, McDonnell PJ, McDonald MB (eds): *Corneal Laser Surgery*. St Louis, Mosby–Year Book 1995, pp 213–226.
17. Talamo JH, Steinert RF, Puliafito CA: Clinical strategies for excimer laser therapeutic keratectomy. *Refract Corneal Surg* 1992;8:319–324.
18. Ohman L, Fagerholm P, Tengroth B: Treatment of recurrent corneal erosions with the excimer laser. *Acta Ophthalmol* 1994;72:461–463.
19. Dausch D, Landesz M, Klein R, Schroder E: Phototherapeutic keratectomy in recurrent corneal epithelial erosion. *Refract Corneal Surg* 1993;9:419–424.
20. Fagerholm P, Fitzsimmons TD, Orndahl M, Ohman L, Tengroth B: Phototherapeutic keratectomy: Long-term results in 166 eyes. *Refract Corneal Surg* 1993;9(Suppl 2):76–81.
21. Forster W, Grewe S, Atzler U, Lunecke C, Busse H: Phototherapeutic keratectomy in corneal diseases. *Refract Corneal Surg* 1993;9(Suppl 2):85–90.
22. Rapuano CJ, Laibson PR: Excimer laser phototherapeutic keratectomy. *CLAO J* 1993;19:235–240.
23. Rapuano CJ, Laibson PR: Excimer laser phototherapeutic keratectomy for anterior corneal pathology. *CLAO J* 1994;20:253–257.
24. John ME, Van der Kass MA, Moblitt RL, Boleyn KL: Excimer laser phototherapeutic keratectomy for treatment of recurrent corneal erosion. *J Cataract Refract Surg* 1994;20: 179–181.
25. Hahn TW, Sah WJ, Kim JH: Phototherapeutic keratectomy in nine eyes with superficial corneal diseases. *Refract Corneal Surg* 1993;9(Suppl 2): 115–118.
26. Campos M, Nielsen S, Szerenyi K, Garbus JJ, McDonnell PJ: Clinical follow-up of phototherapeu-

tic keratectomy for treatment of corneal opacities. *Am J Ophthalmol* 1993;115:433–440.
27. Marshall J, Trokel S, Rothery S, Krueger RR: A comparative study of corneal incisions induced by diamond and steel knives and two ultraviolet radiations from an excimer laser. *Br J Ophthalmol* 1986;70: 482–500.
28. van Setten GB, Koch JW, Tervo K, et al: Expression of tenascin and fibronectin in the rabbit cornea after excimer laser surgery. *Graefes Arch Clin Exp Ophthalmol* 1992;230:178–182.
29. Sanders D: Clinical evaluation of phototherapeutic keratectomy—VisX twenty/twenty excimer laser. FDA submission, written communication, Feb. 7, 1994.
30. Keates RH, Drago PC, Rothchild EJ: Effect of excimer laser on microbiological organisms. *Ophthalmic Surg* 1988;19:715–718.
31. Gottsch JD, Gilbert ML, Goodman DF, Sulewski ME, Dick JD, Stark WJ: Excimer laser ablative treatment of microbial keratitis. *Ophthalmology* 1991;98: 146–149.
32. Pepose JS, Laycock KA, Miller JK, et al: Reactivation of latent herpes simplex visus by excimer laser photokeratectomy. *Am J Ophthalmol* 1992;114: 45–50.
33. Krueger RR, Campos M, Wang XW, Lee M, McDonnell PJ: *Arch Ophthalmol* 1993;111: 1131–1137.
34. McCally RL, Hochheimer BF, Chamon W, Azar DT: A simple device for objective measurement of haze following excimer ablation of cornea. *SPIE* 1993; 1877:20–25.
35. Azar DT: Epithelial and stromal wound healing following excimer laser keratectomy. *Semin Ophthalmol* 1994;9:102–105.
36. Courant D, Fritsch P, Azema A, et al: Corneal wound healing after photo-kerato-mileusis treatment on the primate eye. *Laser Light Ophthalmol* 1990;3: 189–195.
37. Hanna KD, Pouliquen Y, Waring GO III, et al: Corneal stromal wound healing in rabbits after 193-nm excimer laser surface ablation. *Arch Ophthalmol* 1989;107:899–900.
38. Fountain TR, de la Cruz Z, Green WR, Stark WJ, Azar DT: Reassembly of corneal epithelial adhesion structures after excimer laser keratectomy in humans. *Arch Ophthalmol* 1994;112:967–972.
39. Bende T, Seiler T, Wollensak J: Side effects in excimer corneal surgery: Corneal thermal gradients. *Graefes Arch Clin Exp Ophthalmol* 1988;226: 277–280.
40. Vrabec MP, Anderson JA, Rock ME, et al: Electron microscopic findings in a cornea with recurrence of herpes simplex keratitis after excimer laser phototherapeutic keratectomy. *CLAO J* 1994;20:41–44.
41. Chamon W, Azar DT, Stark WJ, Reed C, Enger C: Phototherapeutic keratectomy. *Ophthalmol Clin North Am* 1993;6:399–413.
42. Gartry D, Muir MK, Marshall J: Excimer laser treatment of corneal surface pathology: A laboratory and clinical study. *Br J Ophthalmol* 1991;75:258–269.
43. Seiler T, Bende T, Winckler K, Wollensak J: Side effects in excimer corneal surgery: DNA damage as a result of 193 nm excimer laser radiation. *Graefes Arch Clin Exp Ophthalmol* 1988;226:227–80.
44. Krueger RR, Sliney DH, Trokel SL: Photokeratitis from subablative 193-nanometer excimer laser radition. *Refract Corneal Surg* 1992;8:274–279.
45. Cennamo G, Rosa N, Rosenwasser GOD, Sebastiani A: Phototherapeutic keratectomy in the treatment of Avellino dystrophy. *Ophthalmologica* 1994;208: 198–200.
46. John ME, Martines E, Cvintal T, Ballew C: Excimer laser photoablation of primary familial amyloidosis of the cornea. *Refract Corneal Surg* 1993;9(Suppl 2): 138–141.
47. Fagerholm P, Ohman L, Orndahl M: Phototherapeutic keratectomy in herpes simplex keratitis: Clinical results in 20 patients. *Acta Ophthalmol* 1994; 72: 457–460.
48. Goldstein M, Loewenstein A, Rosner M, Lipshitz I, Lazar M: Phototherapeutic keratectomy in the treatment of corneal scarring from trachoma. *J Refract Corneal Surg* 1994;10(Suppl 2):290–292.
49. McDonnell JM, Garbus JJ, McDonnell PJ: Unsuccessful excimer laser phototherapeutic keratectomy. Clinicopathologic correlation. *Arch Ophthalmol* 1992; 110:977–979.
50. O'Brart DP, Gartry DS, Lohmann CP, et al: Treatment of band keratopathy by excimer laser phototherapeutic keratectomy: Surgical techniques and long-term follow-up. *Br J Ophthalmol* 1993;77: 702–708.
51. Steinert RF, Puliafito CA: Excimer laser phototherapeutic keratectomy for a corneal nodule. *Refract Corneal Surg* 1990;6:352.
52. Moodaley L, Liu C, Woodward GE, O'Bart D, Muir MK, Buckley R: Excimer laser superficial keratectomy for proud nebulae in keratoconus. *Br J Ophthalmol* 1994;78:454–457.
53. Dausch D, Landesz M, Schorder E: Phototherapeutic keratectomy in recurrent corneal intraepithelial dysplasia. *Arch Ophthalmol* 1994;112: 22–23.

CHAPTER 8
Corneal Nodules and Scars

Joshua A. Young, Ernest W. Kornmehl

INTRODUCTION

Corneal nodules and scars are seen in several corneal disorders including Salzman's degeneration, apical scars of keratoconus, corneal partial- and full-thickness lacerations, and anterior adherent leukomas. PTK is valuable only in those conditions where the pathology predominantly involves the anterior one third of the cornea. Visual acuity may be reduced in these conditions because of the opacity, surface irregularity, or associated refractive error. In most situations of corneal nodules and scars, visual impairment results from a combination of these factors; which have to be addressed meticulously during PTK laser surgery.

COMPARISON WITH CONVENTIONAL THERAPY

The treatment of corneal nodules and scars has three clinical goals: to establish a uniform, smooth surface; to remove opacities causing optical degradation; and to minimize any induced refractive change. Traditionally, lamellar keratoplasty has been employed to achieve these goals.

Lamellar keratoplasty involves a substantial commitment on the part of both the surgeon and the patient. It is performed in an operating room and requires peribulbar or retrobulbar anesthesia, adding to the risks of the surgery the possible complications of retrobulbar hemorrhage or penetration of the globe. In addition to the facility and the surgeon's fee, sizable fees include those for the donor tissue and anesthesiologist.[1]

From a therapeutic standpoint, lamellar keratoplasty has several disadvantages as well. Mechanical debridement of lesions on the corneal surface requires a high degree of familiarity unavailable in most practices, in which these lesions present infrequently. Obtaining a uniform, smooth corneal surface of suitable optical quality requires a submillimeter degree of precision difficult to obtain manually.

Whereas submillimeter smoothness is difficult to assess at the operating microscope, the effect of lamellar keratoplasty on the overall curvature of the cornea is impossible to judge at the time of surgery. A deviation in curvature of less than 10% may result in more than 4D of spherical or astigmatic change, and any benefit gained in precise surface smoothing may be lost in substantial anisometropia.

Retreatment options are few and are limited to treatment of recurrence of the initial pathology or complications arising from the graft.[1] Lamellar keratoplasty is ill-suited to repairing irregular astigmatism or spherical changes induced in earlier procedures. If lamellar keratoplasty cannot be repeated, penetrating keratoplasty may be the only available alternative.

In contrast to lamellar keratoplasty, excimer phototherapeutic keratectomy (PTK) is uniquely suited to the removal of superficial corneal lesions. PTK requires less time and involves less expense than lamellar keratoplasty.[1] Unlike lamellar keratoplasty, PTK is performed in a laser suite not requiring the presence of an anesthesiologist. In addition, donor tissue need not be obtained, further reducing the cost of this procedure. PTK is ultimately less invasive, requiring neither retrobulbar nor peribulbar anesthesia.

PTK provides a degree of precision unavailable in lamellar keratoplasty. Once the ablation rate has been determined, uniform ablation may be planned at a micron level. Moreover, the uniformity of this process relieves much of the surgeon's burden of determining iatrogenic corneal surface distortions visually.

Retreatment options and indications are somewhat broader than those for lamellar keratoplasty. PTK may be repeated for recurrence of the initial pathology, as well as for superficial scarring resulting from the photoablation itself. In addition, induced refractive errors may, to a certain extent, be treated as well, using the modified taper technique introduced in Chapter 7, and described in detail later in this chapter. Although excimer photoablation is better suited to refractive correction than lamellar keratoplasty, neither technique is well suited to correction of the hyperopia which often results from these procedures.

The treatment of elevated corneal lesions is particularly challenging. These lesions may be divided into three broad topographic categories: single central lesions, isolated peripheral lesions, and multiple nodules. Nodules and scars may interfere with visual function as a result of opacity and scatter. In addition, elevated lesions may induce irregular astigmatism by changing the anterior corneal contour. This is particularly evident in central elevated lesions.

DIFFICULTIES ARISING FROM PTK

Photoablative treatment of raised nodules poses several challenges not present in the treatment of scars and flat lesions. The first of these is that while PTK removes corneal tissue, it does not, by itself, change corneal contour irregularities. As long as the ablation rate is constant, the excimer laser will simply translate any surface irregularities onto the new surface.[2] Therefore, the elimination of contour irregularities is an active process on the part of the clinician.

The retention of contour irregularities is not limited to epithelial nodules. Many ablation protocols call for epithelial debridement before ablation.[1,3,4] In these instances, irregularities in the contour of the anterior surface of Bowman's layer may themselves be translated onto the new corneal surface.

It is evident that preexisting contour irregularities may be retained in the postablation surface.[2] This is a function of the uniformity of ablation. Irregularities which do not preexist may be induced by nonuniformity of the ablation rate. A corneal surface may be smooth and regular at the beginning of photoablative treatment. But if the lesions to be treated ablate more slowly than the surrounding normal tissue, the corneal contour will change. A slowly ablating lesion will develop into a peak and the surrounding normally ablating tissue into a gully. In this way, the postablation surface will be *less regular* than the initial, pretreatment surface. O'Brart et al.[5] described such a differential in ablation rate in the treatment of rough band keratopathy. The lower ablation rate of the calcific tissue would result in relative elevations and irregular astigmatism if mechanical débridement were not performed first.

Unexpectedly low photoablation rates are not limited to calcific lesions. Dense collagenous scar in the absence of calcification may ablate more slowly than native tissue as well.[6,7] In this way, an initially smooth corneal surface may be made irregular iatrogenically.

To a lesser extent, differential ablation rates may occur in the absence of abnormal lesion ablation. Occasionally, epithelium is left intact at the time of treatment, and ablation is performed through epithelium before reaching stroma. This need not pose a problem if the epithelial layer is uniform. However, epithelium has a higher ablation rate than stroma. A collagenous extension into the epithelium may become amplified by photoablation. As the more slowly ablating collagenous lesion is reached, the surrounding epithelium will ablate at a higher rate, forming a "valley." In truth, the difference between epithelial and stromal ablation rates are sufficiently small to limit this complication to theoretical discussion.

The treatment of central lesions poses a special challenge. Selective ablation of the central cornea, even if performed in a uniform, smooth manner, may result in substantial refractive change. Central flattening leads to hyperopia, a condition not readily amenable to refractive surgical intervention. Hersh et al.[8] reported on a series of 12 eyes treated with PTK for a variety of pathologies, 8 of which experienced substantial hyperopic shifts. Although 3 of the 12 eyes became more myopic (including 2 of two patients with Salzmann's nodules), eight patients experienced hyperopic shifts with a mean change in refractive error of +5.40 D. Similar data have been presented for patients undergoing PTK with the standard taper technique (described below) demonstrating +5.11 D of induced hyperopia at 3 months and 5.28 D at 36 months.[9] An even greater degree of hyperopic shift was reported by Stark et al.[10] in which 58% of patients treated with ablations 85 μm or deeper experienced a change of 9 D or more.

The etiology of this hyperopic shift is twofold. First, many investigators, for reasons described below, choose to treat discrete lesions with small-diameter ablations. If a central lesion is treated this way, the central cornea will be preferentially ablated. This relative depression will result in central flattening and hyperopia similar to that achieved by photorefractive keratectomy (PRK).

A second way in which hyperopia is produced is more insidious. Even if a large-diameter beam is applied uniformly to the entire cornea, hyperopia will still result. This is because the angle of incidence is closer to the normal in the central cornea than in the periphery.[10] Therefore, greater power per square millimeter of corneal surface is applied to the center than to the periphery. The result is disproportionate central ablation, central flattening, and hyperopia. In our experience, differences in angle of incidence for optical radiation are common. One difference between summer and winter is the mean angle of incidence of the sun's rays to the Earth's surface (being greater in winter). An approximate change of 1 D in the direction of hyperopia may be expected for each 20 μm of ablation.[9]

SURFACE MODULATORS

As has already been described, the treatment of an irregular surface by unassisted PTK results merely in the translation of the irregularities to a deeper layer of cornea.[2] Similarly, the unassisted photoablation of a smooth surface consisting of materials with dissimilar ablation rates results in the creation of surface irregularities in the postablation surface.[5-7] Surface irregularities lead to optical degradation regardless of the clarity of the medium. In an effort to reduce preexisting irregularities and to prevent the production of new irregularities, the employment of surface modulators has been introduced.

Surface modulators are materials which influence the ablation of the substrate on which they are applied. These modulators are fashioned into the desired contour and ablated along with stroma to transfer the desired contour to the postablation surface. One such surface modulator consists of a collagen gel molded into the desired contour using the posterior surface of a poly-methyl methacrylate (PMMA) contact lens.[11,12] As the cornea undergoes photoablation, the collagen gel is first exposed to the laser. Later, the "peaks" of the nature tissue are exposed as ablation progresses to their level. Last exposed are the "valleys" of the native tissue, which are ablated only after surrounding tissue has reached their level. This procedure depends upon the collagen gel's completely filling each valley and upon the uniformity of its ablation rate.

The most common surface modulators are fluids which prevent penetration of the excimer radiation. Upon application, these fluids fill depressions in the corneal surface, thereby providing selective masking. As such, they are referred to as *masking fluids*.[2] These have included 0.9% saline (Unisol), 1% carboxymethylcellulose sodium (Celluvisc), and 0.3% hydroxypropylmethylcellulose 2910 with 0.01% dextran 70 (Tears Naturale II). Two important properties distinguish these solutions from each other: absorbance and viscosity. At the critical 193-nm emission wavelength for the argon-fluoride (ArF) excimer wavelength, all three of these solutions seem to have adequate absorbance, with Tears Natural II and Celluvisc showing a somewhat greater absorbance than Unisol. At slightly higher wavelengths, this disparity becomes substantially greater.[2] Although this is of little consequence in ArF excimer photoablation, this difference may be significant in photoablation with newer lasers such as the Novatec which employ slightly longer wavelengths.

The second important property of a masking fluid is its viscosity. Solutions with very low viscosity are believed to be prone to "runoff" or too rapid drainage from surface depressions. Treatment with such masking fluids is assumed to lead to postablation irregularities as a result of inadequate masking after rapid drainage (Figure 8.1). Solutions with very high viscosity are believed to be susceptible to nonuniform layering on the corneal surface, thereby leading to nonuniform masking and an irregular result[2] (Figure 8.2). The ideal viscosity is one which allows adequate retention in depressions while maintaining the ability to coat uniformly (Figure 8.3).

A similar difficulty is presented by the choice of an appropriate laser pulse repetition rate. Lower repetition rates mean longer treatments but have been demonstrated to result in smoother postablation surfaces.[13] It has been postulated that lower

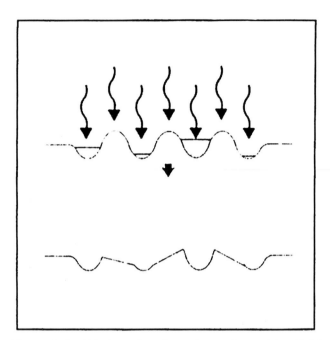

Figure 8.1. Ablation of an irregular surface with a fluid of inadequate viscosity will fail to adequately protect the valleys, in which run-off can occur, and irregularities will persist as ablation progresses. (From Ref. 2, p 861.)

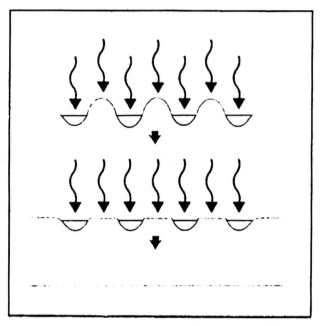

Figure 8.3. The ideal fluid will have adequate absorption and moderate viscosity to result in ablation of the peaks while masking the valleys. (From Ref. 2, p 861.)

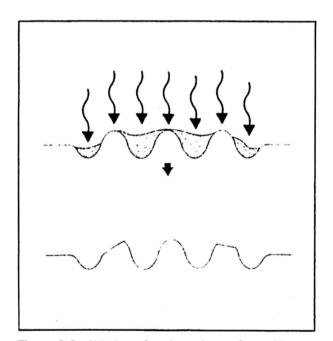

Figure 8.2. Ablation of an irregular surface with a very viscous fluid will be irregular due to the variable coating of different thicknesses of the fluid. (From Ref. 2, p 861.)

repetition rates allow time for masking fluids to refill corneal depressions. One may speculate that masking fluids with higher viscosity may require lower repetition rates than those with lower viscosity.

Surprisingly, surface tension of masking fluids has not been investigated to date. The masking fluid would not be expected to layer in a level fashion between corneal peaks. Rather, a meniscus would likely form, covering the "walls" of some corneal peaks and nonuniformly covering some depressions. The degree of meniscus formation might be modified by the addition of surfactants to the masking fluid.

In some situations, the epithelium itself may be used as a surface modulator. If Bowman's layer is believed to be irregular on examination, a smooth epithelium could serve much the same role as the above-described collagen gel.[9,10] Similarly, surface nodules may be mechanically trimmed back to the epithelial surface before ablation.[1] Subsequent ablation would remove the nodule simultaneously with the epithelium, promoting a uniform result.

SPECIFIC STRATEGIES FOR PTK OF ELEVATED NODULES AND SCARS

Photoablation of corneas with elevated nodules, such as in Salzman's degeneration or optical scars of keratoconus, poses sizable challenges, including an irregular initial contour and a substantial risk of postablation hyperopia. The final goal of therapy is, of course, a smooth postoperative surface. The interim goal, and the one on which we shall now focus, is the creation of a smooth pretreatment surface (Figures 8.4 to 8.7).

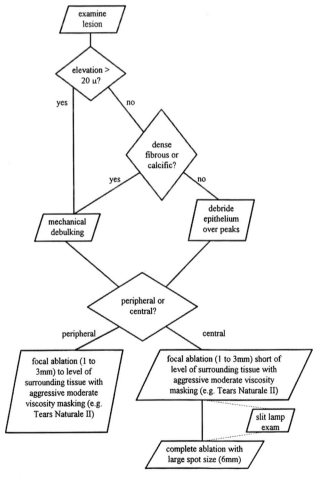

Figure 8.4. Strategies for isolated central or peripheral elevated lesion.

Mechanical Debridement

Several investigators have proposed methods for obtaining a smooth preablation surface in the context of raised nodules. Most of these involve a variable degree of mechanical debridement. Durrie et al.[14] have recommended removing the bulk of the lesion mechanically prior to photoablation. Leveling the lesion preoperatively simplifies the surface modulation necessary to obtain a smooth result.

Hersh et al.[8] recommend removing only epithelium overlying a raised nodule and leaving the surrounding epithelium intact. The remaining epithelium then serves as a surface modulator for the selective ablation of the lesion. A combination of these techniques has been suggested by Talamo et al.[1] in which discrete lesions of elevated more than 20 μm are treated with "preparatory surgical keratectomy" before photoablation. Broad lesions or those elevated less than 20 μm receive no such mechanical débridement and undergo only local epithelial removal as long as no concurrent epithelial pathology is present.

A further modification has recently been proposed by Azar et al.[15] Their method requires the removal of epithelium overlying the peak of an isolated nodule, as well as a small amount of epithelium *surrounding* the peak. This results in the formation of an annular furrow around the elevated lesion. This furrow is then filled with a masking fluid to facilitate selective ablation of the peak. This technique is especially valuable in the management of keretoconus opical scaning, and isolated Salzman's nodules.

Surface Modulators

Although collagen gel surface modulation has shown some promise, the surface modulator of choice for most investigators has been a masking fluid. The most popular masking fluids have moderate viscosity, as these fluids are felt to resist runoff. Many topical lubricating solutions provide a good balance of viscosity and masking. Chief among these Tears is Naturale II. Other masking fluids in clinical use include 0.5%, 1%, and 2% methylcellulose, as well as 1% carboxymethylcellulose sodium (Celluvisc), although this latter fluid has been shown to be less effective in reducing irregularities than the less viscous Tears Naturale II.[2]

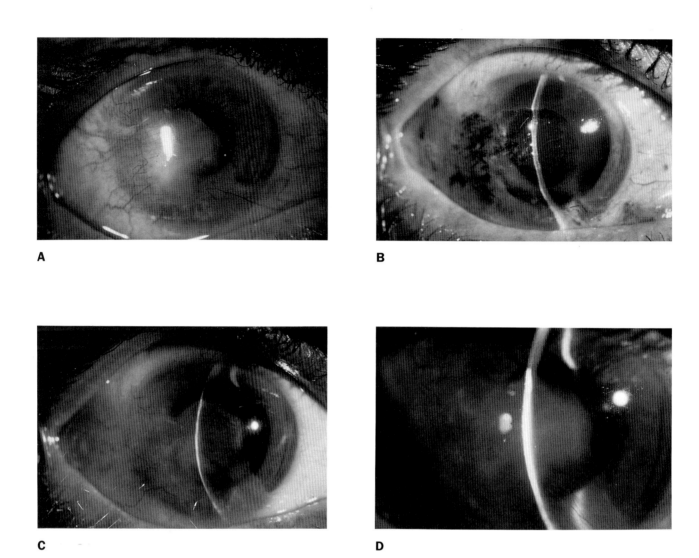

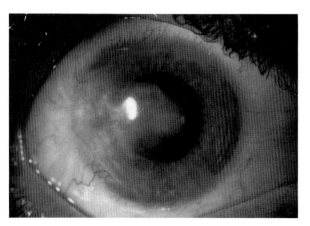

Figure 8.5. Photograph of a patient with corneal scar secondary to recurrent pterygium. *A:* Preoperative appearance. *B:* Intraoperative appearance. Note the difficulties with bleeding. *C:* Two months after PTK. *D:* Nine months after PTK. *E:* Twelve months after PTK.

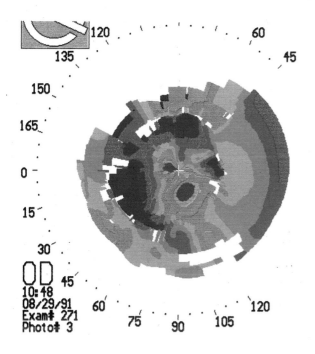
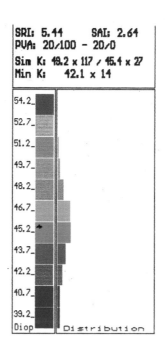

Figure 8.6. Corresponding topography of the patient shown in Figure 8.4 12 months postoperatively.

Thompson et al.[16] recommend a two-tier approach for large, elevated lesions and deep depressions. They begin their ablations with 2% methylcellulose, a relatively highly viscous fluid. Their reasoning is that resistance to runoff of the masking fluid is the most important factor in the early phase of the treatment of a very irregular lesion. Later, as surface regularity improves, the switch to less viscous 1% methylcellulose is made.

As important as choosing the correct masking agent is applying the agent properly. Too thick an application will impede the ablation; too thin, inadequate masking will result in the translation of surface irregularities deeper into the cornea. The masking fluid is applied with a dampened cellulose sponge. The amount applied should be sufficiently thick to fill the depressions, yet not so thick as to cover the peaks. Adequate layering of the masking fluid is evaluated during photoablation using the *sight and sound* method.[16] Many masking fluids, including methylcellulose and 0.3% hydroxypropylmethylcellulose 2910 with 0.01% destran 70, turn white when exposed to excimer radiation. Therefore if, after applying a masking solution and exposing the cornea to the excimer, the peaks appear white, the masking fluid has been applied too thickly. This is the "sight" component. A masking fluid exposed to an excimer pulse emits only a soft "click," whereas naked cornea exposed to the same pulse emits a loud "snap." Therefore, insufficient coverage of a cornea with masking solution results in a "louder" than expected treatment.

Variability of a lesion's ablation rate probably plays less of a role in the ablation of raised nodules than it does in the removal of other pathology such as calcific band keratopathy. Masking fluids would be expected to aid in the prevention of iatrogenic irregularities resulting from the treatment of such low ablation rate areas. One might expect collagen gel surface modulation to be less effective in such settings. Collagen gel surface modulation is predicated on the gel's ablation rate equaling that of the substrate. If a dense, fibrous or calcific lesion is overlaid with collagen gel, the differences in ablation rate between the lesion and the surface modulator will result in the formation of a peak. However, unlike a masking fluid, the collagen gel will not be able to flow into the iatrogenically developed valleys and an irregular surface will result. Once the collagen solution has gelled, it cannot reconfigure itself to compensate for developing irregularities. In contrast, masking fluids can be redistributed during

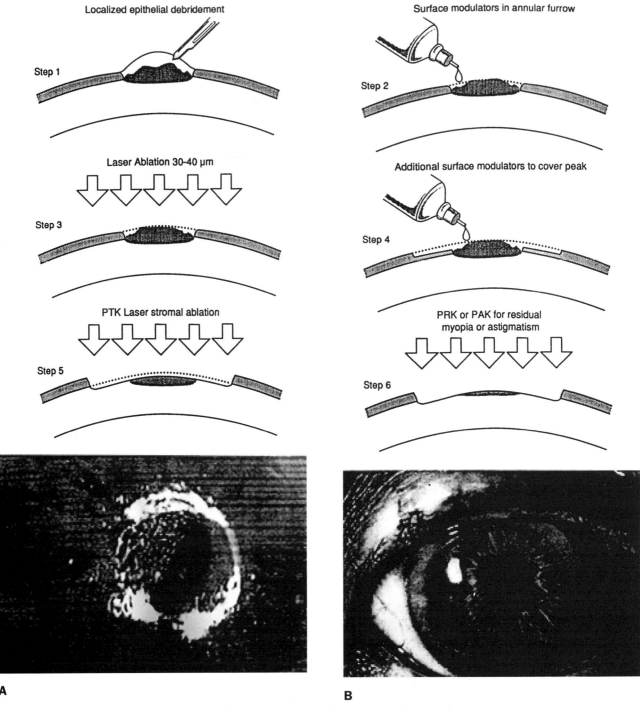

Figure 8.7. Surgical technique of PTK for elevated central corneal nodules. Steps 1–6 are schematically illustrated in top drawings. *A:* Intraoperative appearance of the central elevated nodule is shown with irregular epithelium overlying nodule prior to epithelial scraping and application of surface modulators. *B:* Smooth corneal surface after PTK. The linear elevated opacity noted preoperatively had disappeared. (Reproduced from Azar DT, Jain S, Stark W: Phototherapeutic Keratectomy. In *Refractive Surgery* Azar DT (ed.). Appleton and Lange, Stanford CT, 1996, pp. 504 and 513).

ablation. In this way, the collagen gel may be described as a static modulator as opposed to masking fluids, which are dynamic modulators.

Ablation Strategies

The primary goal of the surgeon performing PTK should be to apply the most conservative ablation necessary to achieve a good result. Although some investigators[10] have treated Salzmann's nodules with large spot sizes (ablation diameters), many investigators recommend focal treatment of discrete elevated lesions. Talamo et al.[1] describe the treatment of a dense, fibroblastic, elevated nodule with a 1-mm ablation diameter. A more recent description by Azar[15] reports a good result after ablating a central Salzmann's nodule with a 3-mm treatment zone (Figure 8.7). Earlier focal ablation techniques, referred to as *point and shoot*, produced poor results, chiefly because of abrupt edges between treated and untreated cornea.[14] These sharp drop-offs tended to lead to corneal fibrosis and epithelial hypertrophy. The use of surface modulators has made focal ablation safer and spares nondiseased cornea from excimer exposure.

More widely distributed multiple nodules may not allow the surgeon to perform focal treatment. If nodules are few in number and widely distributed, they may be approached in the same way as isolated nodules. If, on the other hand, the nodules are densely packed, a larger ablation zone may be selected, with close attention paid to adequate masking.

After an appropriate ablation spot size is chosen, a pulse repetition rate must be selected. Although the clinical evidence is sparse, a slower (lower-frequency) repetition rate is probably preferable to a faster (higher-frequency) one. A commonly chosen repetition rate is 10 MHz, but investigation has suggested that slower rates (as low as 2 Hz) may produce smoother results.[13] Explanations proposed for this difference include improved redistribution of the masking agent in the longer interpulse interval of the 2-Hz rate and dissipation of airborne debris liberated from the corneal surface.

In addition to producing a smooth, uniform surface, the surgeon wishes to minimize iatrogenic hyperopia. This is achieved in several ways. The VISX 20/20 excimer laser allows the production of a *standard taper*, a transition zone 0.5 mm in width, providing a gradual blending from ablated to untreated zones. This is generally believed to be inadequate in focal ablations, and several other techniques has been developed. One early attempt to blend the ablation edge involved rotating the eye in a circular manner while delivering the beam through a varying aperture size. This technique was later abandoned because ablation depth proved difficult to judge.[9] A more widespread technique involves placing the surgeon's hands to the sides of the patient's head and briskly[14] or gently[15] moving the head in a circular manner beneath the beam. A 2-mm spot size is chosen for this procedure, and ablation is performed to a depth of 20 μm.[9,10] This manual blending, known as *modified taper*, achieves two goals. First, it produces a smooth transition from untreated to fully ablated tissue. Second, the annular pattern thus produced may help to counteract the central flattening which is the source of iatrogenic hyperopia.

Azar et al[15] have modified this technique further by performing an additional 6-mm ablation over a partial-thickness 3-mm focal ablation. The nodule is first debulked, but not completely removed, with a 3-mm focal treatment. Leveling of the residual lesion is completed with the larger 6-mm spot size, thereby producing a multistep ablation (Figure 8.7).

SUGGESTED PROTOCOLS

Isolated Central or Peripheral Elevated Nodule or Scar

The first step in treatment is to examine the lesion. Any elevations of excessive height (defined as 20 μm or more beyond the epithelial surface) will require mechanical debulking before ablation to avoid extensive excimer exposure, which may lead to irregular astigmatism and hyperopia. Next, attempt to determine the density and composition of the lesion. Pay special attention to the presence of calcium. Dense fibrous and calcific lesions may exhibit a retarded ablation rate. In this way, a lesion which is minimally elevated preoperatively may result in a larger peak after treatment. Particularly dense or calcific lesions should be mechanically débrided before ablation is performed. If debridement is not performed, epithelium overlying the

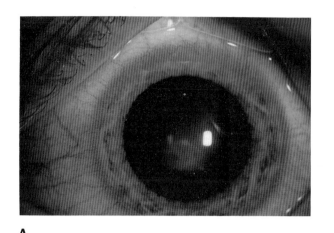

A

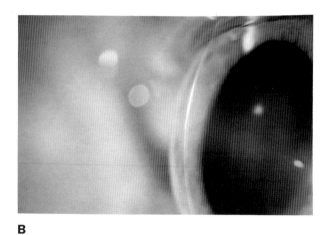

B

C

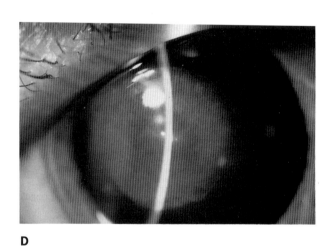

D

E

Figure 8.8. A 33-year-old man with depressed corneal scar presenting with 20/30 vision and grade 1+ mean haze who underwent PTK. *A, B:* Clinical appearance of the cornea prior to PTK. Note the depression in the area of scarring. *C:* Appearance 2 months after surgery. *D, E:* Appearance 5 months postoperatively by direct illumination and retroillumination, respectively.

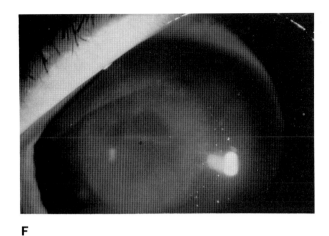 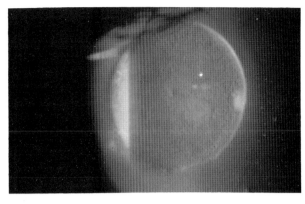

Figure 8.8. *F, G:* Appearance 9 months postoperatively by direct illumination and retroillumination, respectively. Note the worsening of scarring if the treated area was depressed originally rather than elevated. Visual acuity remained at 20/30 despite the worsening of scarring.

peak should be removed. This epithelium may be removed in such a way as to create an annular depression in which the masking fluid may pool.

Next, an appropriate ablation diameter must be chosen. Choose a spot size (1 to 3 mm) which will be large enough to encompass the lesion. Ablation is performed while the patient's head is gently moved in a circular pattern. The patient, however, continues to fixate on the target light (i.e., straight ahead). If the lesion is peripheral, ablation is performed to the level of the surrounding untreated tissue. If the lesion is central, focal ablation is terminated just shy of the surrounding level. A 6-mm or larger ablation zone is then chosen to complete the leveling process. Aggressive fluid masking with Tears Naturale II or 1% methylcellulose is performed in either case. Appropriate masking is assessed with the sight and sound technique described above. Ablation is terminated when a level surface is obtained, keeping in mind that undertreatment is preferable to overtreatment (Figures 8.4 and 8.9).

Multiple Elevated Nodules or Scars

Again, the first step in treatment is a careful assessment of the lesions. If they are few in number and sparse in distribution, the above technique may be applied. The finding of many or closely spaced raised lesions renders focal treatment impractical. Unless the elevations are relatively minor, some degree of mechanical debridement is probably warranted. As with single nodules, the presence of a dense scar or calcification should sway the surgeon toward mechanical debridement as well.

The key to treatment of multiple lesions is adequate masking. Initial irregularities are masked with 2% methylcellulose for the first half of the ablation sequence. Subsequent ablation is performed with Tears Naturale II or 1% methylcellulose. Consideration may be given to decreasing the pulse repetition rate in order to maximize smoothing. A large 6-mm ablation zone is chosen, and no circular head movements need be performed. However, after adequate ablation has been achieved, an additional shallow annular ablation is performed. A 2-mm spot size is chosen and, by moving the patient's head, is made to trace the edge of the ablation to a depth of 20 μm (Figures 8.10 to 8.12).

Postoperative Management

Following PTK, efforts concentrate on promoting reepithelialization. Immediately postoperatively, a drop of cycloplegic solution and an antibiotic-corticosteroid ointment are applied, followed by a pressure patch. Oral analgesics are prescribed. The patient is seen 24 h later, and the patch is removed. An antibiotic-corticosteroid combination is continued until reepithelialization is complete, at which

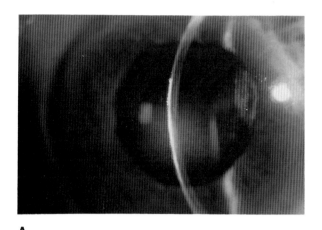

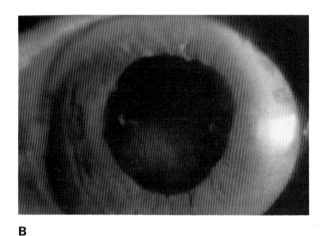

Figure 8.9. *A:* Preoperative appearance of a 85-year-old woman with corneal scar and astigmatism, resulting in visual acuity of 20/100. *B:* One year after PTK, with 20/70 visual acuity.

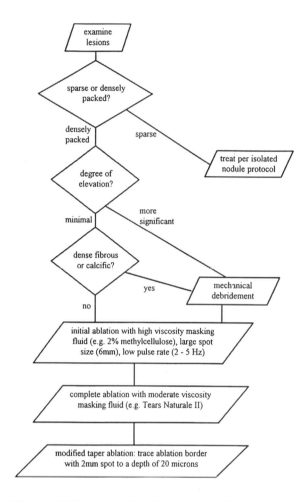

Figure 8.10. Strategies for multiple elevated lesions.

Corneal Nodules and Scars

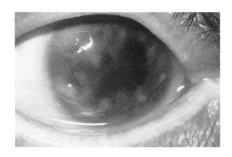

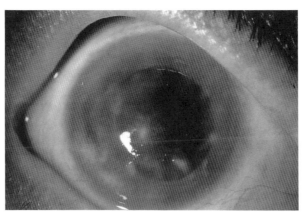

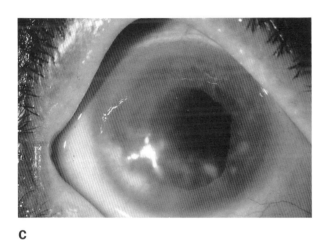

Figure 8.11. A 77-year-old man with multiple peripheral elevated Salzmann's nodules in the right eye who presented with 20/1400 vision and a 2+ mean haze score. *A:* Preoperative phtotograph of the eye showing the lesion. *B:* Appearance 3 months after PTK. *C:* Nine months after PTK. The mean haze score improved to zero, but the visual acuity worsened to 20/2000.

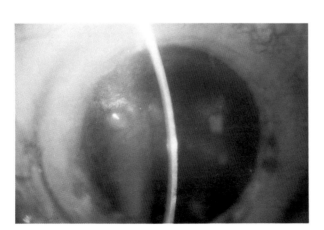

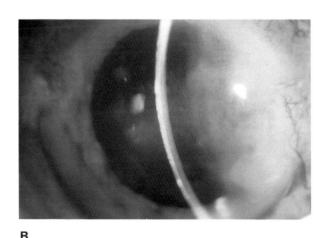

Figure 8.12. *A, B:* Slit lamp photographs of a 70-year-old man with multiple elevated Salzmann's nodules in the left eye resulting in visual acuity of 20/200.

(continued)

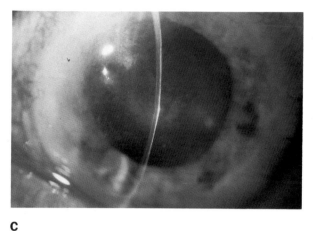

C

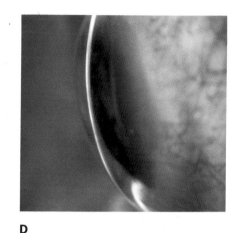

D

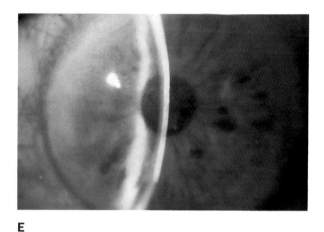

E

Figure 8.12. *C:* Appearance 3 months after PTK. Note the smoothness of the central nodule compared to the neighboring superonasal nodule. *D:* Appearance 9 months after surgery, with 20/100 visual acuity. *E:* Appearance 12 months postoperatively. Visual acuity of 20/70 was achieved at that time.

time the antibiotic is discontinued. The topical corticosteroid is tapered over a period of several months.

REFERENCES

1. Talamo JH, Steinert RF, Puliafito CA: Clinical strategies for excimer laser phototherapeutic keratectomy. *Refract Corneal Surg* 1992;8:319–324.
2. Kornmehl EW, Steinert RF, Puliafito CA: A comparative study of masking fluids for excimer laser phototherapeutic keratectomy. *Arch Ophthalmol* 1991; 109:860–863.
3. Rogers C, Cohen P, Lawless M: Phototherapeutic keratectomy for Reis-Bucklers' corneal dystrophy. *Aust NZ J Ophthalmol* 1993;21:247–250.
4. Campos M, Nielsen S, Szerenyi K, Garbus JJ, McDonnell PJ: Clinical follow-up of phototherapeutic keratectomy for treatment of corneal opacities. *Am J Ophthalmol* 1993;115:433–440.
5. O'Brart DPS, Gartry DS, Lohmann CP, Patmore AL, Kerr Muir MG, Marshall J: Treatment of band keratopathy by excimer laser phototherapeutic keratectomy: Surgical techniques and long term follow-up. *Br J Ophthalmol* 1993;77:702–708.
6. McDonnell JM, Garbus JJ, McDonnell PJ: Unsuccessful excimer laser phototherapeutic keratectomy. *Arch Ophthalmol* 1992;110:977–979.
7. Binder PS, Anderson JA, Rock ME, Vrabec MP: Human excimer keratectomy. *Ophthalmology* 1994; 101:979–989.
8. Hersh PS, Spinak A, Garrana R, Mayers M: Phototherapeutic keratectomy: Strategies and results

in 12 eyes. *Refract Corneal Surg* 1993;9(Suppl): S90–S95.
9. Azar DT, Jain S, Woods K, Stark W, Sanders DR, Kinoshita S: Phototherapeutic keratectomy: The VisX experience. *Corneal Laser Surg* Salz J (ed) Mosby, 1995.
10. Stark WJ, Chamon W, Kamp MT, Enger CL, Rencs EV, Gottsh JD: Clinical follow-up of 193 nm ArF excimer laser photokeratectomy. *Ophthalmology* 1992; 99:805–812.
11. Englanoff JS, Kolahdouz-Isfahani AH, Moreira H, et al: In situ collagen gel mold as an aid in excimer laser superficial keratectomy. *Ophthalmology* 1992;99: 1201–1208.
12. De Vore DP, Scott JB, Norquist RE, Hoffman RS, Nquyen H, Eiferman RA: Rapidly polymerized collagen gel as a smoothing agent in excimer laser photoablation. *J Refract Surg* 1995;11:50–55.
13. Fasano AP, Moreira H, McDonnell PJ, Sinbawy A: Excimer laser smoothing of a reproducible model of anterior corneal surface irregularity. *Ophthalmology* 1991;98:1782–1785.
14. Durrie DS, Schumer DJ, Cavanaugh TB: Phototherapeutic keratectomy: The summit experience. *Corneal Laser Surg* Salz J (ed) Mosby, 1995.
15. Azar DT, Jain S, Stark W: Phototherapeutic keratectomy. In *Refractive Surgery* Azar DT (ed). Appleton and Lange, Stamford CT, 1996, pp. 502–511.
16. Thompson V, Durrie DS, Cavanaugh TB: Philosophy and technique for excimer laser phototherapeutic keratectomy. *Refract Corneal Surg* 1993;9(Suppl): S81–S85.

CHAPTER 9
Recurrent Erosion Syndrome

Suhas W. Tuli, Dimitri T. Azar, Walter J. Stark, Perry S. Binder

INTRODUCTION

Recurrent erosions were first described in 1872 by Hansen.[1] These are seen in a variety of corneal erosive disorders due to dystrophies involving the epithelium, stroma, and endothelium; physical, chemical, thermal, or infectious insults to the cornea; and miscellaneous causes such as diabetes mellitus, neurotrophic keratitis, and keratitis sicca. Current standard therapy may induce patching, 5% NaCl, cycloplegia, topical steroids, bandage contact lenses, epithelial debridement, anterior stromal puncture, and yitrium-aluminum-garnet (YAG) laser surgery.[2–7] Although most eyes respond well to conventional treatment, some become recalcitrant, and treatment of these eyes can be frustrating to the patient and the physician.

The pathogenesis of recurrent erosion syndrome (RES) is defective adhesion between the epithelium and the stroma due to abnormalities in adhesion structures, including hemidesmosomes, basement membrane, and anchoring fibrils.[2,8,9] Phototherapeutic keratectomy (PTK) with 193-nm excimer laser has been reported to be an effective treatment of recurrent erosions.[10–16] The premise for using PTK to treat RES is that smoothing of the subepithelial surface enhances epithelial migration and adhesion in a manner similar to that of superficial keratectomy (Table 7.2.). PTK has also been used to treat several corneal pathological disorders including superficial stromal dystrophies and stromal scarring.[10,11,17–22] (See Chapters 7 and 8)

PTK FOR RECURRENT EROSIONS

Like any new technique, however, PTK itself has the potential to cause serious complications. The reformation of epithelial and stromal adhesion structures is delayed following PTK.[23–27] Clinically, recurrent erosions have been reported in patients following excimer keratectomy.[28–30] It is therefore unclear whether PTK should be used only as a last recourse in the treatment of RES.

Chapter 13 describes PTK clinical outcomes using the Summit Excimer for the treatment of corneal disorders including epithelial basement membrane dystrophy (65 eyes) and recurrent erosions (64 eyes). The results were very favorable. However it is not clear from the published data whether patients with recalcitrant RES benefited from PTK for extended periods of time (Table 7.2.).

PTK was performed in five eyes of four patients with recurrent corneal epithelial erosions following trauma and map-dot-fingerprint dystrophy. These patients had undergone standard medical and surgical treatment for recurrent corneal erosions but proved recalcitrant to therapy (Table 9.1).

Patients were treated with the VISX 20/20 excimer laser (Santa Clara, CA; pulse rate of 5 Hz and fluence of 160 mJ/cm^2). All patients met the eligibility criteria approved by the U.S. Food and Drug Administration (FDA), were informed of the investigational nature of the laser procedure, and provided written informed consent prior to surgery. Preoperative and postoperative evaluations were performed on all patients, including external, slit lamp, and dilated fundus examinations. Visual acuity was evaluated without correction, with manifest refraction, with pinhole, and with a hard contact lens. When possible, pneumotonometry, keratometry, ultrasonic and optical pachymetry, and topographic analysis were performed.

Results

The four clinical patients (one man and three women) undergoing PTK had a mean age of 36.75

Table 9.1. Clinical Outcome in Four Patients Following PTK

Patient No.	Age/Sex/Eye	Diagnosis	Recurrences/Prior Treatment	Preoperative Data		PTK	
				V/A	Refraction	Depth	Diameter
(1)	36/f/OS	Cogan's dystrophy	1 every 2 weeks/epi. débridement, bandage CL	20/20	−0.25 −1 × 175	30 μm	6 mm
(2)	47/m/OD	Recurrent erosions, SNVM	lubrication	20/300	1.25 −0.75 × 146	8 μm	6 mm
(3)	31/f/OD	Recurrent erosion with acne rosacea	Almost daily epi. débridement stromal puncture	20/50	0.00	5 μm	6 mm
(4)	33/f/OD	Traumatic recurrent erosions	Almost daily/micropuncture, bandage CL	20/40	−0.50 −1.00 × 173	First t/t 3 μm, periphery 10 μm	3 mm / 2 mm
(4)	33/f/OD	Recurrent erosions	—	20/50	−0.75 −0.75 × 160	Second t/t 8 μm	6 mm
(4)	33/f/OS	Traumatic recurrent erosions	Almost daily/micropuncture, epi. débridement, bandage CL	20/50	−0.25 −0.50 × 150	First t/t 3 μm, inferiorly 1 μm	6 mm / 2 mm
(4)	33/f/OS	Recurrent erosions	—	20/40	−0.50 −0.25 × 180	Second t/t 13 μm, outside prior treated area	2 mm

Abbreviations: V/A, corrected visual acuity; CL, contact lens; SNVM, subretinal neovascular membrane; epi, epithelial; t/t, treatment.

years (range, 13–47 years). The mean follow-up period was 19.4 months (range, 12 to 37 months). The results are summarized in Table 9.1.

Patients 1 and 2 responded very well to PTK and had no recurrences for 12 months following treatment. Patients 3 and 4 suffered recurrences after PTK. The right eye and left eye of patient 4 were recurrence free for only 3 and 1 months, respectively, and required retreatment. Despite repeat PTK, recurrences continued in both eyes. The success rate of PTK for RES is lower than that reported in other studies.[6,10–14] The best corrected visual acuity improved by one Snellen line or more in all four cases. None of the eyes showed worsening of visual acuity or a significant hyperopic shift. PTK used to treat RES, with success rates ranging from 74% to 100%.[10–16] We considered PTK treatment only in patients who were recalcitrant to all other medical and surgical therapies. It is noteworthy that the patient with poor results after the first PTK did not do well in spite of retreatment. In addition, there are reports of delayed epithelialization in eyes undergoing PTK.[10,32] Gartry et al.[30] have reported epithelial instability in 3% of eyes undergoing PTK. Map-fingerprint-dot changes in the corneal epithelial basement membrane have been reported following radial keratotomy (RK) and excimer laser keratectomy.[28,29,33] These findings indicate that the histopathological alterations following excimer keratectomy may be clinically relevant and should be taken into consideration when contemplating PTK for treating recurrent corneal erosions.

HISTOPATHOLOGICAL FINDINGS

Morphometric Measurement of Basement Membrane Zone Adhesion Structures

Ten different patients had undergone previous excimer laser treatment for a variety of corneal lesions, including corneal dystrophies, healed viral keratitis, and scarring (Table 9.2). Surface irregu-

3-Month Follow-up			6-month Follow-up			12-Month Follow-up		
V/A	Refraction	Recurrences	V/A	Refraction	Recurrences	V/A	Refraction	Recurrences
20/15	−0.50 −0.50 × 170	None	20/15	−1.00 +0.50 × 85	None	20/15	—	None
20/80	1.00 −1.00 × 12	None	20/60	0.25 −1.25 × 15	None	20/70	0.25 −1.50 × 171	None
20/25	−0.75 −0.5 × 15	Moderate	20/40	0.00	Mild	20/60		Moderate
20/25	−0.75 −1.00 × 165	None	20/20	−0.75 −0.75 × 162	Mild	20/40	−0.50 −1.00 × 155	Moderate
20/25	−0.25 −1.25 × 175	Severe	—	—	—	—	—	—
20/50	0 −0.75 × 180	Moderate outside treated area	20/40	−0.25 −0.25 × 5	Moderate	—	—	—
20/40	−1.75 −0.50 × 8	—	20/30	−0.75 −0.50 × 175	None	20/25	−1.50	Moderate

larities or subepithelial scarring had developed postoperatively that were significant enough to warrant corneal transplantation, which was performed 5–16 months after excimer treatment. Corneal buttons from three different areas of each wound bed, as previously described,[26,27] were examined with electron micrographs. The micrographs were digitized using a Dage-MTI Image digitizer, as reported previously[23,34] (Dage-MTI, Inc., Michigan City, IN; Apple Macintosh 11Cx Workstation, IMAGE version 1.42 Rasband, 1992). The following three parameters were measured: (1) the length of the basal cell membrane, (2) the cross-sectional area of the basal lamina, and (3) the length of the basal cell membrane covered by hemidesmosomes. The percentage of basal cell membrane occupied by hemidesmosomes [3/1 × 100] and cross-sectional area of basal lamina per 100 μm of basal cell membrane [2/1 × 100] were calculated. Also were numbered the basal epithelial cells and underlying stromal keratocytes in micrographs recorded at magnifications of 22,000× or 36,000×. This was used to determine the percentage of basal epithelial cells overlying multilaminated and discontinuous basal laminas; the percentage of cells overlying normal anchoring fibrils, hemidesmosomes, and irregular collagen; and the percentage of epithelial basal cells within 2.5 μm of a keratocyte.

The results of electron microscopic analysis in the second series of treated corneas and the clinical findings are shown in Table 9.2. The hemidesmosome counts were normal as early as 5 months after excimer laser keratectomy. The average percentage of basal cell membrane covered with hemidesmosomes was also normal at that time. Most of the corneal specimens had discontinuous basal lamina 5–11 months after excimer treatment. However, the 15- to 16-month samples showed normal and multilaminated basal lamina. The percentage of cells with multilaminated basal lamina increased from 12% at 6 months to 73% at 16 months. The percentage of cells with normal anchoring fibrils increased pro-

Table 9.2. Reassembly of Adhesion Structures in 10 Patients Undergoing PTK

Case No.	Diagnosis	Pre-PTK V/A	Intended PTK Depth (μm)	Interval to PK After PTK (mo)	Pre-PK V/A	Indication for PK
(a)	Adenoviral subepithelial opacity	20/50	15*	5	—	Central scar
(b)	Macular dystrophy	20/70	50*	6	20/50	Residual dystrophy
(c)	Granular dystrophy	—	140*	7	—	Residual dystrophy
(d)	Posttraumatic scar	20/50	130	8	—	Residual scar
(e)	Keratoconus	20/100	71*	9	20/70	Subepithelial central scar, myopic shift
(f)	Recurrent lattice dystrophy	—	30*	10	—	residual dystrophy, and scar
(g)	Stromal scar status after bacterial keratitis	20/60	113*	11	20/50	Central scar, hyperopic shift
(h)	Band keratopathy	20/300	120*	15	20/100	Central scar, hyperopic shift
(i)	Herpes simplex keratitis scar	20/50	200*	16	—	Central scar
(j)	Herpes simplex keratitis scar	20/400	100	16	—	Recurrence of herpes simplex and diffuse scarring

Abbreviations: PTK, phototherapeutic keratectomy; V/A, corrected visual acuity; PK, penetrating keratoplasty; BL, basal lamina; HD, hemidesmosomes.
*Epithelium débrided prior to PTK.
†Mean percentage of basal cell membrane covered with hemidesmosomes.
‡Mean cross-sectional area of basal lamina per 100 μm of basal cell membrane.

gressively over time (Figure 9.1). The percentage of epithelial basal cells with 2.5 μm of a stromal keratocyte decreased progressively from 5 months to 11 months, while the number of cells overlying irregular collagen increased from 17% at 5 months to 61% at 11 months (Figure 9.2). Histopathological findings after PTK indicate that the epithelial-stromal reassembly of adhesion structures is delayed, with permanent alterations in architecture.[23–27,35] Discontinuity in the basal lamina, disorganized lamellar structure of the subepithelial tissue, decrease in the number of cells with normal anchoring fibrils, and increase in the cross-sectional area of the basal lamina per 100 μm of basal cell membrane have been reported. Although our electron microscope findings are in conformity with those of previous studies, there are several limitations of our study, including (1) potential misinterpretation of the qualitative data secondary to shrinkage artifacts, magnification errors, and fixation and processing techniques, and (2) the fact that the corneal buttons were obtained from patients who had undergone treatment for various corneal pathologies that may have contributed to the alterations in architecture.

PTK OPERATIVE TECHNIQUES FOR RES

Preoperatively, the patients should be alerted regarding possible complications including infection, pain, recurrences in or around the area of treatment, hyperopic shift, and failure to achieve desired effects of the procedure. Oral analgesics and topical anesthetics are usually administered prior to surgery. PTK can be performed during or after an episode of recurrent erosions. Invariably, the epithelial surface is irregular and the basal epithelial cells are poorly adherent to the stroma. The epithelium can be scraped easily with a nonfragmenting sponge (such as Weckcel or MuroCel). In fact the epithelium may come off as a sheet extending beyond the areas of recurrent defects. Mechanical removal of all loose epithelium should be perfromed prior to PTK laser treatment. With rare exceptions, the laser treatment should be cen-

	Percentage of cells with:		% HD/BCM[†]	BL/100 μm BCM[‡]
Normal HD	Multilaminate BL	Discontinuous BL		
70	40	50	32	25
59	12	59	39	17
30	4	85	16	11
92	58	33	30	40
66	26	58	26	22
90	40	30	29	24
80	10	52	27	20
81	3	24	29	24
23	63	17	15	33
79	73	12	35	39

tered around the pupil using a 6.0 or 6.5 mm-diameter zone. If this zone covers the areas of recurrent defects, and if the epithelium has been completely scraped in the treatment zone, a 3–5 μm treatment depth is performed after sequential application of wet and dry nonfragmenting sponges over the stromal surface.

If the treatment zone overlaps an area of intact epithelium, a blade or spatula is used to scrape the epithelium prior to treatment. Subsequent localized peripheral laser treatments may be also used to treat areas of recurrent erosions outside the Central treatment zone. Again, the laser treatment depth should be limited to 3–5 μm.

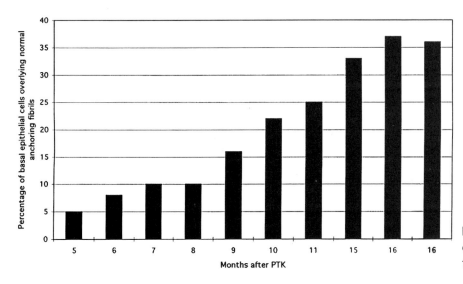

Figure 9.1. Percentage of basal epithelial cells with normal anchoring fibrils at different times after PTK.

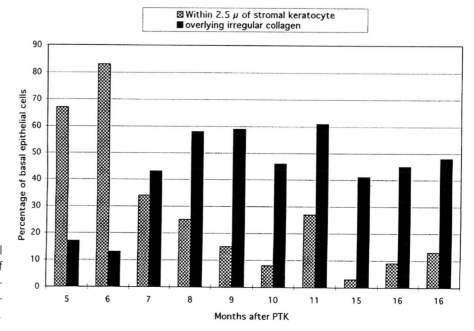

Figure 9.2. Percentage of basal epithelial cells with 2.5 μm of stromal keratocytes and percentage of cells overlying irregular collagen at different times after PTK.

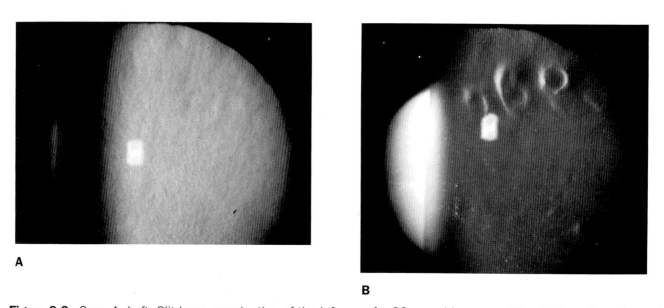

Figure 9.3. Case 1. Left: Slit lamp examination of the left eye of a 36-year-old woman with a history of recalcitrant recurrent erosions prior to PTK reveals corneal erosions at the 4:30 and 8:00 positions close to the limbus. Right: Slit lamp examination 1 year postoperatively shows no evidence of erosions. The depth and diameter of treatment were 30 μm and 6 mm, respectively. The patient was recurrence free at this time.

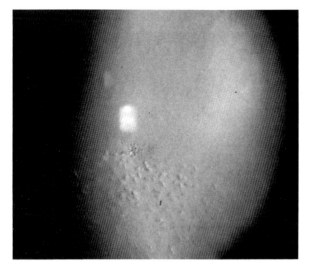

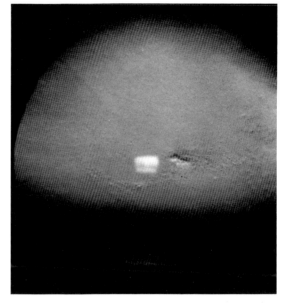

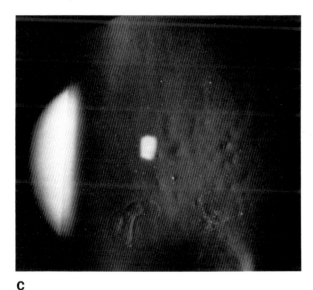

Figure 9.4. Case 4. Top: Slit lamp examination of a 33-year-old woman with recalcitrant recurrent erosions in the left eye prior to PTK shows erosions centrally and stromal puncture inferiorly. Middle: 6 months following the first PTK (3 μm, 6 mm centrally and 1 μm, 2 mm inferiorly), the same eye revealed erosions centrally on slit lamp examination. Bottom: Two years after the second PTK (13 μm, 2 mm), the patient experienced recurrences in spite of retreatment. Slit lamp examination showed erosions centrally and inferiorly.

SUMMARY

Histopathological alterations in the basement membrane zone can be observed as late as 16 months after PTK (Table 9.2). Morphometric measurements revealed abnormalities including a high percentage of cells with discontinuous basal lamina and a decreased number of cells with normal anchoring fibrils. These findings are similar to those reported after traumatic, diabetic, and dystrophic RES.[2,8,9,36,37] Furthermore, we clinically observed recurrence of erosions in three of six eyes following treatment of recalcitrant RES with PTK. We presume that the documented histological abnormalities in the basement membrane zone may be linked to clinical recurrences after PTK.

We believe that PTK should not be used to treat patients with untreated or early recurrent corneal erosions. Given (1) indications from previous reports that the stromal architecture does not stabilize even by 16 months, (2) our electron microscopic findings of alterations in the epithelial basement membrane and adhesion structures following PTK, and (3) the potential for recurrence after PTK, we therefore recommend alternative treatment for such patients with RES. Most patients respond well to topical lubrication therapy, bandage soft contact lenses, debridement of epithelium or basement membrane, or anterior stromal micropuncture.[24]

PTK may still have a beneficial role in patients with recalcitrant RES. The Summit data (Chapter 13) has shown favorable outcomes, specifically reduced pain, improved comfort, and improved vision.

Because PTK has the potential to cause a hyperopic shift and other serious complications, careful case selection and meticulous surgical technique, as well as analysis of therapeutic goals are essential for its success.

REFERENCES

1. Hansen E: Om den intermitterende keratitis vesiculosa neuralga af traumatisk oprindsele. *Hospital Stidende* 1872;15:201–203.
2. Kenyon KR: Recurrent corneal erosion: Pathogenesis and therapy. *Int Ophthalmol Clin* 1979;19:169–195.
3. McLean EN, MacRae SM, Rich LF: Recurrent erosion: Treatment by anterior stromal puncture. *Ophthalmology* 1986;93:784–786.
4. Wood TO: Recurrent erosion. *Trans Am Ophthalmol Soc* 1984;82:895–898.
5. Katz HR, Snyder ME, Green WR, Kaplan HJ, Abrams DA: Nd:YAG laser photo-induced adhesion of the corneal epithelium. *Am J Ophthalmol* 1994;118:612–622.
6. Azar DT, Jain S, Woods K, Stark W, Sanders D: Photoherapeutic keratectomy: The VISX experience. In Salz JJ (ed): *Corenal Laser Surgery*. Hanover, MD, Mosby–Year Book, 1994, pp 148–164.
7. Buxton JN, Fox ML: Superficial epithelial keratectomy in the treatment of epithelial basement membrane dystrophy; a preliminary report. *Arch Ophthalmol* 1983;101:392–395.
8. Cogan DG, Kuwabara T, Donaldson DD: Microcystic dystrophy of the cornea. *Arch Ophthalmol* 1974;92:468–474.
9. Rodrigues MM, Fine BS, Laibson PR, Zimmerman LE: Disorders of the corneal epithelium. *Arch Ophthalmol* 1974;92:475–482.
10. Sher NA, Bowers RA, Zabel RW, et al: Clinical use of the 193-nm excimer laser in the treatment of corneal scars. *Arch Ophthalmol* 1991;109:491–498.
11. Forster W, Grewe S, Atzler U, Lunecke C, Russe H: Phototherapeutic keratectomy in corneal diseases. *Refract Corneal Surg* 1993;9:S85–S90.
12. Fagerholm P, Fitzsimmons TD, Orndahl M, Ohman L, Tengroth B: Phototherapeutic keratectomy: Long term results in 166 eyes. *Refract Corneal Surg* 1993;9(Suppl 2):76–81.
13. Rapuano CJ, Laibson PR: Excimer laser phototherapeutic keratectomy for anterior corneal pathology. *CLAO J* 1994;20:253–257.
14. John ME, Van der Karr MA, Noblitt RL, Boleyn KL: Excimer laser phototherapeutic keratectomy for treatment of recurrent corneal erosion. *J Cataract Refract Surg* 1994;20:179–181.
15. Dausch Landesz M, Klein R, Schroder: Phototherapeutic keratectomy in recurrent corneal epithelial erosion. *Refract Corneal Surg* 1993;9:419–424.
16. Ohman L, Fagerholm P, Tengroth B: Treatment of recurrent corneal erosions with the excimer laser. *Acta Ophthalmol* 1994;72:461–463.
17. Chamon W, Azar DT, Stark WJ, Reed C, Enger CL: Phototherapeutic keratectomy. *Ophthalmol Clin North Am* 1993;6:399–413.
18. Hahn TW, Sah WJ, Kim JH: Phototherapeutic keratectomy in nine eyes with superficial corneal diseases. *Refract Corneal Surg* 1993;9(Suppl 2):S115–S118.
19. Talamo JH, Steinert RF, Puliafito CA: Clinical strategies for excimer laser therapeutic keratectomy. *Refract Corneal Surg* 1992;8:319–324.
20. Steinert RF: Therapeutic keratectomy: Corneal smoothing. In Thompson FB, McDonnell PJ (eds): *Color Atlas/Text of Excimer Laser Surgery: The Cornea*. New York and Tokyo, Igaku-Shoin, 1993, pp 121–129.
21. McDonnell PJ, Seiler T: Phototherapeutic keratectomy with excimer laser for Reis-Bücklers' corneal dystrophy. *Refract Corneal Surg* 1992;8:306–310.
22. Steinert RF, Puliafito CA: Excimer laser phototherapeutic keratectomy for a corneal nodule. *Refract Corneal Surg* 1990;6:352.
23. Fountain TR, de la Cruz Z, Green WR, Stark WJ, Azar DT: Reassembly of corneal epithelial adhesion structures after excimer laser keratectomy in humans. *Arch Ophthalmol* 1994;112:967–972.
24. Vrabec MD, Anderson JA, Rock ME, et al: Electron microscopic findings in a cornea with recurrence of herpes simplex keratitis after excimer laser

phototherapeutic keratectomy. *CLAO J* 1994;20: 41–44.
25. McDonnell JM, Garbus JJ, McDonnell PJ: Unsuccessful excimer laser phototherapeutic keratectomy. Clinicopathologic correlation. *Arch Ophthalmol* 1992;110:977–979.
26. Wu WCS, Stark WJ, Green WR: Corneal wound healing after 193-nm laser keratectomy. *Arch Ophthalmol* 1991;109:1426–1432.
27. Binder PS, Anderson JA, Rock ME, Vrabec MP: Human excimer laser keratectomy. Clinical and histopathologic correlations. *Ophthalmology* 1994; 101:979–989.
28. Busin M, Meller D: Corneal epithelial dots following excimer laser phototherapeutic keratectomy. *Refract Corneal Surg* 1994;10:357–359.
29. Maguen E, Salz JJ, Nesburn AB, et al: Results of excimer laser phototherapeutic keratectomy for the correction of myopia. *Ophthalmology* 1994;101: 1548–1556.
30. Gartry DS, Kerr-Muir MG, Marshall J: Efficacy and long-term complications of excimer laser photorefractive keratectomy (PRK)—3 year follow-up. *Invest Ophthalmol Vis Sci* 1993(Suppl);34:892.
31. Uozato H, Guyton DL: Centering corneal surgical procedures. *Am J Ophthalmol* 1987;103:264–275.
32. Stark WJ, Chamon W, Kamp MT, Enger CL, Rencs EV, Gottsch JD: Clinical follow-up of 193-nm ArF excimer laser photokeratectomy. *Ophthalmology* 1992;99:805–811.
33. Nelson JD, Williams P, Lindstrom RL, Doughman DJ: Map-fingerprint-dot changes in the corneal epithelial basement membrane following radial keratotomy. *Ophthalmology* 1985;92:199–205.
34. Freund DE, McCally RL, Goldfinger AD, Farrell RA: Image processing of electron micrographs for light scattering calculations. *Cornea* 1993;12: 466–474.
35. Krueger R: Wound healing after keratorefractive surgery. In Salz JJ (ed): *Corneal Laser Surgery.* Hanover, MD, Mosby–Year Book, 1994, pp 12–48.
36. Azar DT, Spurr Michaud SJ, Tisale AS, Gipson IK: Decreased penetration of anchoring fibrils into diabetic stroma: Morphometric analysis. *Arch Ophthalmol* 1989;107:1520–1524.
37. Azar DT, Spurr Michaud SJ, Tisdale AS, Gipson IK: Altered epithelial basement-membrane interactions in diabetic cornea. *Arch Ophthalmol* 1992;110: 1468–1471.

CHAPTER 10
PTK Complications

Marco C. Helena, Jonathan H. Talamo

INTRODUCTION

As discussed in previous chapters, the 193-nm argon-fluoride excimer laser ablates the corneal tissue with submicron precision, causing little damage to surrounding, nonexposed areas.[1-3] Its principal applications include (1) correction of refractive errors (PRK) and (2) treatment of a variety of anterior corneal pathologies (PTK). In PTK different strategies are employed according to the type and extension of the pathology and the therapeutic goal in each particular case. Although general treatment guidelines for some pathologies have been proposed,[4-6] the surgical technique may vary, depending not only on the patient's corneal status but also on the skill, experience, and judgment of the surgeon. We describe the most frequent complications associated with PTK and propose principles for their prevention and treatment.

REFRACTIVE CHANGES

As with any corneal surgery, a smooth, regular anterior corneal surface following PTK is required for a better visual outcome. The corneal epithelium plays a crucial role in modulating such smoothness, with its ability to compensate for a certain degree of stromal irregularity (Figure 10.1). However, major changes in the anterior stromal profile may significantly alter the refractive characteristics of the corneal surface, resulting in irregular astigmatism as well as myopic or hyperopic shifts.

Irregular Astigmatism

Irregular astigmatism is a refractive error that cannot be corrected with sphero-cylindrical spectacle lenses. Corneal surface abnormalities frequently lead to irregular astigmatism, which may arise from small, focal abnormalities or diffuse areas of topographic irregularity on visually significant portions of the cornea. Various PTK components may also result in irregular astigmatism.

Epithelial Removal

In general, the corneal epithelium has a regular surface despite the presence of minor anterior stromal irregularities, playing the role of a natural masking and smoothing agent as tissue is ablated. In most cases of PTK, a transepithelial ablation is performed. However, epithelial removal is indicated when the epithelial surface is irregular to avoid reproducing such topography on the stromal surface. Mechanical epithelial removal with a metal blade is the most frequently used technique. In some cases, however, the epithelium is strongly attached to the underlying corneal layers, and its removal is difficult. Adjunctive chemical deepithelialization techniques (e.g., ethylene diamine tetra acetate (EDTA) in band keratopathy) may occasionally be helpful.

Varying Ablation Rates Among Tissues

The ablation rates of different corneal tissue constituents by the excimer laser have been extensively studied.[7-11] When the energy density (fluence) of 193-nm excimer laser light exceeds a critical value, individual molecular bonds are irreversibly broken, leading to tissue ablation. Different molecular structures have different ablation rates; consequently, so do different tissue components. The ablation rate of the epithelium is the highest, followed in descending order by stroma, Bowman's layer, corneal scars, and calcium deposits. If excimer laser energy is applied directly to a surface where tissues with different ablation rates coexist, such as calcium deposits in band keratopathy or

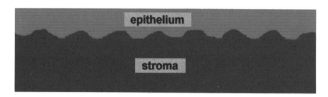

Figure 10.1. The epithelium compensates for a certain degree of stromal irregularity, providing a smooth anterior corneal surface.

stromal scars of any etiology, the resulting treatment surface may be irregular (Figure 10.2A). In these cases, the use of masking fluids is indicated during ablation. Depending on their viscosity and surface tension, masking fluids are able to fill in depressions and expose elevations of an irregular corneal surface. Because they absorb laser energy, they can shield tissue components with higher ablation rates while overexposing tissues with lower ablation rates (Figure 10.2B). As discussed in Chapters 7 and 8, a variety of substances have been used for this purpose. By comparing the smoothing effects of 0.3% hydroxypropyl-methylcellulose 2910 and 0.1% dextran 70 solution (Tears Naturale II), 1% carboxymethylcellulose sodium (Celluvisc), and 0.9% saline (Unisol) on irregular corneal surfaces, Kornmehl et al.[13] obtained better results in the Tears Naturale II treatment group.

Setting of Ablation Zone Parameters

Paracentral anterior stromal lesions are common indications for PTK, as they may decrease visual acuity by blocking the entrance of light rays or by causing irregular astigmatism, as well as glare and light scattering. Although treating each individual lesion with an appropriately sized ablation zone diameter set according to the size of the lesion can potentially decrease the total amount of tissue ablation, asymmetric paracentral ablations may induce significant irregular astigmatism. Glare and/or monocular diplopia may also be induced if the margins of paracentral ablations are too close to the center of the entrance pupil. Alternatively, paracentral lesions may be treated with large optical zones centered over the entrance pupil, using masking agents to shield normal corneal tissue as necessary.

Hyperopic Shift

The PTK mode of photoablation was designed to ablate corneal tissue uniformly without altering surface refractive power. Thus, the surface profile of

Figure 10.2. The role of masking fluids in PTK.

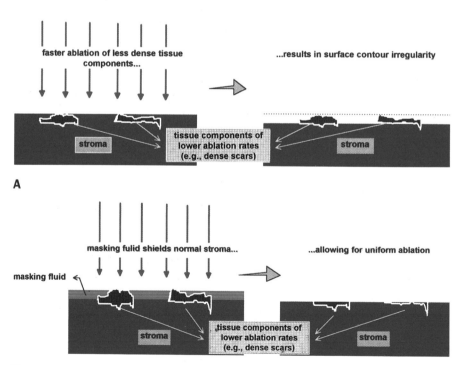

a standard polymethyl methacrylate test block is preserved after applying the excimer laser with a PTK setting. However, following PTK in human corneas, hyperopic shifts have frequently been observed.[6,12,14-19] In a series of 35 PTKs, Sher et al.[14] observed hyperopic shift in 50%. Campos et al.[17] detected hyperopic shift in 56% of 18 consecutive cases. Potential mechanisms for the hyperopic shift are as follows: (1) ablation products may provide greater shielding toward the edge of the ablation zone if deposited in a centrifugal fashion; (2) the decreasing angle of incidence of the laser beam on more peripheral cornea may reduce laser efficacy as the edge of the ablation zone is approached; and (3) the abrupt margins of the ablation zone following PTK may result in peripheral epithelial and stromal hyperplasia, creating a myopic lens effect[20] (Figure 10.3).

Strategies to reduce the hyperopic shift have been developed. Sher et al.[14] initially described a smoothing technique in which the eye was moved via head rotation in a circular manner under a laser beam of varying aperture size. Since it was impossible to predict depth measurements accurately, this technique was abandoned. Later, the same authors decided to perform a hyperopic ablation immediately following an initial therapeutic ablation. Stark et al.[15] attempted to decrease central flattening by moving the eye under the laser beam and treating the margins of the ablation zone with a 20-μm-deep, 2-mm-diameter spot size. A reduction in hyperopic shift was achieved. Sanders et al.[21]

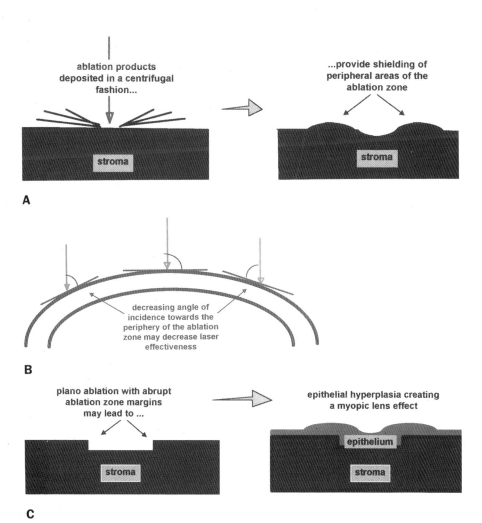

Figure 10.3. Hyperopic shift after PTK.

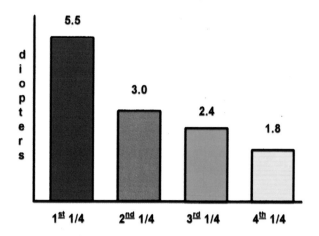

Figure 10.4. Sanders et al.[21] divided 271 patients undergoing PTK at the VisX centers into four groups, or quartiles, based on the data of treatment. The mean degree of hyperopic shift decreased progressively from the first (5.5 D) to the last quartile (1.8 D).

divided the refractive changes of 271 patients who underwent PTK at the U.S. VISX centers into quartiles according to the dates of treatment. The degree of postoperative hyperopic shift was larger in the first quartile and was progressively reduced as more refined techniques were developed (Figure 10.4).

A positive correlation between depth of ablation and hyperopic shift has been detected.[16,18] Therefore, monitoring the depth of ablation required in each case is extremely important to avoid unnecessary tissue removal. When removing stromal opacities, it is often difficult to assess ablation depth under the operating microscope. It is very helpful to remove tissue conservatively, frequently returning the patient to an upright position to allow for intraoperative slit lamp examination. It is certainly wiser to be conservative and add more treatment as necessary than to overtreat, which is irreversible.

Myopic Shift

To induce a myopic shift, the cornea has to be more deeply ablated in the periphery than in the center (Figure 10.5). Attempts to correct hyperopia with this technique using the excimer laser have met with limited success, but myopic shifts have occurred following PTK. Sher et al[14] observed myopic shift in 3% and Campos et al.[17] in 16.6% of patients in their series. The exact mechanism of this finding is not clear. Theoretically, if the treatment is directed to paracentral areas, with the central cornea protected by masking agents, such an effect could be achieved.[22]

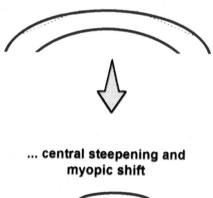

Figure 10.5. Effect of treatment location on refractive outcome.

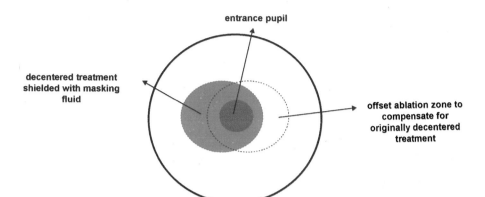

Figure 10.6. Management of decentered excimer laser treatment.

DECENTRATION

Currently, most surgeons believe that the optical zone should be centered around the center of the entrance pupil when performing keratorefractive procedures.[23] Using this technique, Seiler et al.[24] observed decentration of more than 1 mm in 1.1% of their PRK series. Refractive astigmatism in this subgroup increased to 1.5 D at 1 year and decreased to 1.0 D at 2 years, accompanied by an undercorrection of approximately 1.0 D. In 23% of the treated eyes, the eccentricity of the ablation zone was between 0.5 and 1.0 mm, with no impact on refractive astigmatism. Spadea et al.[25] observed decentration of more than 0.5 mm in 30% of myopic PRK patients and suggested that undercorrection can be related, at least in part, to decentration.

Centration of PTK is variable, as it depends on the distribution of surgical pathology. In cases in which extensive treatment is required, the ablation zone should be large and centered, as in PRK. However, treatment of small lesions is generally centered over the lesions themselves, with the laser spot diameter adjusted according to the size of the lesion. When such lesions are peripheral or paracentral, surgeons should proceed with caution, as treatment can result in irregular astigmatism, myopic shift, or monocular diplopia.

Technical factors that may induce decentration include (1) inappropriate patient fixation, (2) misalignment of the laser beam and the microscope, and (3) use of miotics before surgery, which may nasally displace the center of the physiological entrance pupil. Treatment should always be interrupted as soon as poor centration is recognized.

Once adequate centration is obtained, ablation should be resumed after ensuring an appropriate degree of corneal stromal hydration and reapplication of a masking agent if indicated.

Management of decentration is difficult and quite variable. If myopic refraction occurs after a decentered PTK, retreatment may be performed in a stepwise fashion using principles similar to those employed for decentered PRK.[26] See Chapter 12. It may be necessary to shield part of the ablated surface with masking fluids for all or part of the retreatment procedures (Figure 10.6).

INCOMPLETE TREATMENT

Dense corneal scar tissues and calcific deposits are difficult to photoablate. As a result, larger than normal numbers of pulses for a given ablation depth are necessary. Meticulous use of masking fluids is required to avoid extensive ablation of adjacent normal corneal tissue within the ablation zone (Figure 10.7).

An unsuccessful treatment of stromal scarring due to its low ablation rate has been reported.[26] Although some surgeons prefer not to treat band keratopathy by PTK for the same reason, O'Brart et al.[12] obtained good results in a series of 122 eyes. In cases of rough band keratopathy, mechanical keratectomy to remove calcific plaques was performed, followed by PTK to remove fine depositions and smooth the stromal surface. The same approach has been suggested when treating other pathologies, such as Salzmann's nodular degeneration, where again, mechanical removal of the lesion should be

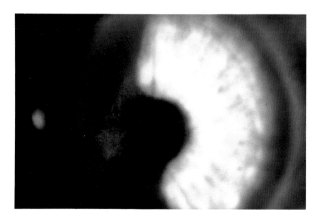

Figure 10.7. Despite the attempt to remove the corneal scar completely, residual superficial opacity persisted following PTK. The area involving the visual axis cleared, however, and the patient required no further treatment.

performed prior to PTK.[5] Such manual "debulking" procedures minimize the need for excessive photoablation of these areas of abnormal corneal tissue, where the tissue ablation rate may fluctuate greatly even within lesions that have a uniform clinical appearance. Since it is difficult to evaluate the depth of treatment through the operating microscope, careful slit lamp examination during surgery may be instrumental in planning and performing PTK.[14]

ENDOTHELIAL CELL DAMAGE

The 193-nm excimer laser produces precise etching of corneal tissue, with minimal damage to adjacent areas. It is completely absorbed by the ablated corneal tissue without transmission to deeper layers, and direct damage to endothelial cells by the excimer laser radiation should not be expected. However, endothelial damage can result from indirect effects. Photoablative procedures modify corneal anatomy and physiology. The corneal stroma is reduced in thickness, and Bowman's membrane is frequently removed. Theoretically, these modifications could affect endothelial metabolism.[27] In addition, the shock waves produced by the impact of each laser pulse can lead to acute mechanical cell trauma.[28]

Endothelial injury has been demonstrated in animal models following excimer laser ablation. Marshall et al.[3] performed deep excimer laser incisions in rabbit corneas and found endothelial cell loss when the ablation depth was within 40 μm of Descemet's membrane. Dehm et al.[30] performed excimer laser and diamond knife incisions to 90% of corneal depth in rabbits and observed similar endothelial alterations with both methods.

The experiments mentioned above, however, do not represent the photoablative procedures currently performed in human beings. Photorefractive keratectomy (PRK) and PTK involve portions of the cornea that are much more superficial. PRK studies have failed to show endothelial damage. Fantes et al.[31] did not detect any endothelial abnormalities by light and electron transmission miscroscopy following PRK in monkeys. Amano et al.[32] performed endothelial specular microscopy in a series of 26 eyes undergoing PRK and observed no statistically significant difference in mean cell density or coefficient of variation of mean cell area among preoperative, 1-month, and 1-year postoperative measurements. In a series 14 human eyes, Perez-Santonja et al.[33] observed no change in endothelial cell density and hexagonality 1 year following PRK, but found significant improvement from preoperative readings of the coefficient of endothelial cell area variation (polymegathism). Carones et al.[28] performed a similar study in 76 eyes. After 1 year, not only did the coefficient of cell area variation improve, but also the percentage of hexagonal cells compared with preoperative values, without significant change in cell density. The authors postulated that the improvement in the endothelial cell parameters postoperatively resulted from discontinuation of contact lens wear.

Endothelial damage is a serious concern following the application of any new corneal surgical technique. At present, there is no indication of clinically significant endothelial damage following PRK or PTK. However, corneal surgeons should remain mindful of the possibility as more long-term follow-up experience accumulates.

PAIN

The human cornea is innervated by the ciliary nerves, which originate from the nasociliary branch of the trigeminal nerve. Sensory impulses are mediated by A-delta-fibers and C-fibers that course from the periphery to the center of the cornea, involving the epithelium and stroma. Following excimer laser ablation, a large number of corneal nerve fibers are intercepted, resulting in pain of variable intensity. Generally, the pain begins 30 to 60 min after the procedure and may become severe within 4 to 6 h, diminishing steadily as reepithelialization occurs.

Biological mediators involved in inflammation and pain following photoablation include arachidonic acid metabolites such as prostaglandins and leukotrienes. Arachidonic acid is released into the intracellular space from cell plasma membranes by phospholipases. Prostaglandins and thromboxanes are generated via the cyclooxygenase pathway, while leukotrienes arise through the lipoxygenase pathway. Postaglandins may increase nerve ending sensitivity and produce hyperalgesia. Leukotrienes may increase pain by promoting inflammatory cell chemotaxis. Phillips et al.[34] demonstrated a sudden and sustained rise in prostaglandin E_2 but no detectable change in leukotriene B_4 in rabbit corneas following PRK.

Initially, management of pain after excimer laser corneal ablation was not satisfactory for a majority of patients. It consisted of counseling and treatment with pressure patching or bandage contact lenses, cycloplegics, ice packing, and oral analgesics including narcotics. The introduction of topical nonsteroidal anti-inflammatory drugs has greatly improved this situation. In general, these drugs reduce the production of prostaglandins and thromboxanes by inhibiting the cyclooxygenase pathway. Because of lipoxygenase pathway overload, however, some of these drugs can enhance leukotriene production, leading to increased inflammatory cell infiltration.[34]

Although topical nonsteroidal anti-inflammatory drugs such as ketorolac tromethamine and flurbiprofen sodium have been shown to be effective in reducing pain after PRK,[35] diclofenac sodium is the agent most extensively tested for corneal excimer laser procedures. Diclofenac has been shown to reduce pain and light sensitivity significantly following PRK.[35,36] Philips et al.[34] observed significant reduction of prostaglandin E_2 but increased leukocyte infiltration using diclofenac in PRK-treated rabbit corneas. However, no detectable change in the level of leukotriene B_4, a potent chemotactic agent, was found. It has been shown[37] that high doses of diclofenac not only inhibit the cyclooxygenase pathway but also decrease the bioavailability of intracellular arachidonic acid. Pretreatment with diclofenac did not produce any additional effect on suppression of prostaglandin E_2 levels following PRK.[38] In addition, Szerenyi et al.[39] demonstrated that diclofenac acutely decreases corneal sensitivity when 1 drop is applied every 5 min to normal, unoperated human eyes. Sensitivity returns to normal within less than 1 h after discontinuation of the drug. The mechanism involved is unclear. Despite claims that diclofenac may delay reepithelialization, Sher et al.[37] found no significant difference in the reepithelialization rate between patients receiving topical diclofenac QID and controls following PRK. In a survey of Canadian surgeons performing PRK, 28 reports (approximately 1 in 250 treated eyes) of subepithelial corneal infiltrates were found in association with pre- and postoperative use of topical nonsteroidal anti-inflammatory drugs (diclofenac or ketorolac) and contact lens fitting (unpublished data, Patricia Teal, M.D., American Society of Cataract and Refractive Surgery meeting, Boston, April 12, 1994). No patient in this series received topical steroids. Sher et al. reported 1 patient with similar corneal infiltrates and 1 patient with an immune ring among 16 patients treated with topical diclofenac QID, as well as 1 patient with a corneal infiltrate among 16 patients treated with placebo. However, in a review of 700 consecutive cases of PRK in which either ketorolac or diclofenac was used postoperatively in conjunction with 0.1% fluorometholone and a bandage contact lens, no complications (including corneal infiltrates) were observed in association with this regimen.[41]

Topical steroids are potent anti-inflammatory drugs, stabilizing lysozomal membranes, inhibiting both arachidonic acid pathways, and decreasing inflammatory cell margination and migration. Fluorometholone has decreased intraocular penetration and may be preferable to minimize topical steroid-induced complications such as glaucoma, cataract, and ptosis. Campos et al.[42] demonstrated that fluorometholone significantly reduces leukocyte infiltration in rabbit corneas 24 h after PRK. Hence, it is possible that fluorometholone may decrease the incidence of corneal infiltrates associated with topical nonsteroidal anti-inflammatory drugs by reducing the influx of inflammatory cells. Since steroidal and nonsteroidal anti-inflammatory drugs have synergistic effects, topical regimens combining both drugs are commonly used following excimer laser procedures.

Although transient hypoesthesia occurs, by 3 months following PRK corneal sensitivity returns to preoperative levels. No complications such as recurrent epithelial defects are associated with this transient reduction in corneal sensitivity.

PERSISTENT EPITHELIAL DEFECTS AND RECURRENT EPITHELIAL EROSIONS:

Rapid, stable reepithelialization after PTK is desirable for a number of reasons, including the reduction of pain, discomfort, inflammation, and risk of infection, as well as the more rapid improvement in visual acuity. In most cases, the epithelium heals completely within 1 week. However, sporadic cases of persistent epithelial defect[6,18,35] or recurrent epithelial erosions[12] have been reported following PTK. The absence of or damage to the remaining Bowman's membrane, the toxicity of topical postoperative medications, particularly non-steroidal anti-inflamatory agents, and the presence of preoperative, active ocular surface and lid inflammatory processes (which can potentiate postoperative inflammation) are possible etiologies. Prevention and management of impaired epithelial healing consist of (1) meticulous preoperative control of the ocular surface and adnexal disease; (2) patching or bandage contact lens fitting; (3) proper corneal lubrication; and (4) suspension or substitution of potentially toxic topical medications. Seiler and Wollensak[44] have reported a patient with systemic lupus erythematosus who developed a persistent epithelial defect which progressed to a sterile corneal ulceration and perforation 30 days after PRK. The authors concluded that autoimmune diseases are a clear contraindication to PRK. The same can be inferred for PTK.

HERPES SIMPLEX VIRUS (HSV) REACTIVATION

HSV is the most common infectious cause of corneal blindness in the Western Hemisphere, and herpetic keratitis is the third most common preoperative diagnosis in patients undergoing penetrating keratoplasty. Within 48 h of primary infection, virions may travel via any of the major divisions of the trigeminal nerve to the trigeminal, ciliary, and sympathetic ganglia, as well as to mesencephalic nuclei of the brain stem, where they enter a latent state. There is also evidence that the cornea may serve as a site of latency. Virus shedding and reactivation may occur. Once reactivated, the virus may travel to the eye and cause infectious blepharoconjunctivitis, keratitis, and keratouveitis. Among the many factors that may trigger HSV reactivation, of special concern for corneal laser surgeons are ultraviolet light, trauma, and immunosuppression.[45]

Currently, the most frequently employed surgical treatment for herpetic corneal scars is penetrating keratoplasty. Drawbacks of this technique include (1) the risks of a major intraocular surgery; (2) the increased risk of graft rejection in inflamed and/or highly vascularized corneas; and (3) the frequent need for intensive topical steroid therapy postoperatively. PTK is a potential alternative for treatment of superficial herpetic scars, with decreased side effects and cost. However, it is possible that HSV reactivation may be induced by the excimer laser. McDonnell et al.[46] described one case of herpes epithelial keratitis recurrence 4 weeks following astigmatic PRK in a patient who had undergone penetrating keratoplasty for herpetic stromal scar. Vrabec et al.[47] reported recurrence of herpetic dendritic keratitis 3 months (2 cases) and 18 months (one case) following PTK to treat stromal scar secondary to recurrent herpetic keratitis. Pepose et al.[48] demonstrated that excimer laser photoablation can reactivate HSV epithelial keratitis in latently infected mice. In addition, there is a report of one case of disciform keratitis which recurred three times in a 28-month period following PTK.[16] Therefore, pre- and postoperative prophylaxis with oral antiviral agents (acyclovir, famcyclovir, or valcyclovir) appropriate when performing PTK on patients with a previous history of HSV blepharoconjunctivitis or keratitis.

GRAFT REJECTION

The main cause of corneal transplant failure is immune-mediated graft rejection occurring in 16–30% of recipients. Cellular immune mechanisms appear to play a key role, but humoral components have also been detected. Risk factors include (1) corneal vascularization, (2) graft size and proximity to the limbus, (3) infections, (4) persistent epithelial

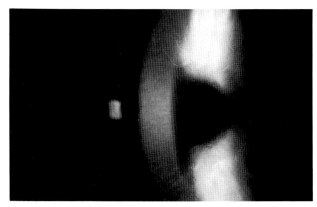

A

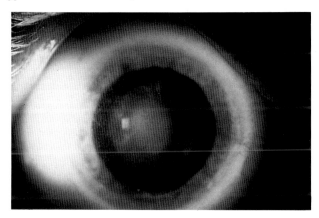

C

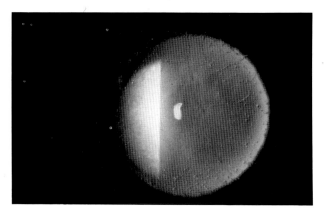

B

Figure 10.8. A 30-year-old woman with lattice dystrophy resulting in 20/200 visual acuity and 2+ mean haze who underwent PTK with deep ablation. *A:* Preoperative slit lamp photograph demonstrating lattice dystrophy with haze. *B:* Appearance 1 month after deep ablation. The treatment resulted in a clear area with visual acuity of 20/70 and a mean haze score of zero. *C:* Clinical appearance 18 months after PTK. Note the subepithelial haze in the area of deep treatment. Visual acuity of 20/60 was achieved in spite of the haze.

defects, and (5) other disorders associated with inflammation, including surgical trauma.[49]

Graft rejection has been reported in two cases following excimer photoablation. One patient underwent PTK to treat recurrent lattice dystrophy in the corneal graft,[50] and another underwent PRK to reduce post-keratoplasty astigmatism.[51] In both cases, rejection was successfully treated medically. Epstein et al.[51] have altered their protocol to include the use of 1% prednisolone every 2 h as opposed to 0.1% fluorometholone following excimer photoablation of corneal grafts.

SUBEPITHELIAL SCARRING

Subepithelial scarring often follows abnormal collagen deposition in the area of treatment (Figure 10.8). The scarring peaks 3–6 months after surgery and subsides thereafter in most patients.

SUMMARY

Excimer PTK is not and should not be a standardized procedure. Strategies and techniques vary, depending upon the surgeon's experience and the individual characteristics of each case. Although a number of different complications may occur, they are rarely sight-threatening. Given the existing alternatives, PTK appears to be a safe and useful technique for the treatment of anterior corneal pathology.

REFERENCES

1. Trokel SL, Srinivasan R, Braren B: Excimer laser surgery of the cornea. *Am J Ophthalmol* 1983;96:710–715.
2. Puliafito CA, Steinert RF, Deutsch TF, et al: Excimer laser ablation of the cornea and lens: Experimental studies. *Ophthalmology* 1985;92:741–748.

3. Marshall J, Trokel S, Rother S, et al: An ultrastructural study of corneal incisions induced by an excimer laser at 193 nm. *Ophthalmology* 1985;92:749–758.
4. Talamo JH, Steinert RF, Puliafito CA: Clinical strategies for excimer laser phototherapeutic keratectomy. *J Refract Corneal Surg* 1992;8:319–324.
5. Thompson V, Durrie DS, Cavanaugh TB: Philosophy and technique for excimer laser phototherapeutic keratectomy. *J Refract Corneal Surg* (Suppl) 1993;9:81–85.
6. Hersh PS, Spinak A, Garrana R, Mayers M: Phototherapeutic keratectomy: Strategies and results in 12 eyes. *J Refract Corneal Surg* (Suppl) 1993;9: 90–95.
7. Krueger SL, Trokel SL: Quantitation of corneal ablation by ultraviolet laser light. *Arch Ophthalmol* 1985; 103:1741–1742.
8. Puliafito CA, Wong K, Steinert RF: Quantitative and ultrastructural studies of excimer laser ablation of the cornea at 193 and 248 nanometers. *Lasers Surg Med* 1987;7:155–159.
9. Van Saarlos PP, Constable IJ: Bovine corneal stroma ablation rate with 193-nm excimer laser radiation: Quantitative measurement. *J Refract Corneal Surg* 1990;6:424–429.
10. Seiler T, Kriegerowski M, Schnoy N, et al: Ablation rate of human corneal epithelium and Bowman's layer with the excimer laser (193 nm). *J Refract Corneal Surg* 1990;6:99–102.
11. Salz JJ, Maguen E, Nesburn AB, et al: A two-year experience with excimer laser photorefractive keratectomy for myopia. *Ophthalmology* 1992;100:873–882.
12. O'Brart DPS, Gartry DS, Lohmann CP, Patmore AL, Kerr Muir MG, Marshall J: Treatment of band keratopathy by excimer laser phototherapeutic keratectomy: Surgical techniques and long term follow-up. *Br J Ophthalmol* 1993;7:702–708.
13. Kornmehl EW, Steinert RF, Puliafito CA: A comparative study of masking fluids for excimer laser phototherapeutic keratectomy. *Arch Ophthalmol* 1991; 109:860–863.
14. Sher NA, Bowers RA, Zabel RW, et al: Clinical use of the 193-nm excimer laser in the treatment of corneal scars. *Arch Ophthalmol* 1991;109:491–498.
15. Stark WJ, Chamon W, Kamp MT, et al: Clinical follow-up of 193-nm ArF excimer laser photokeratectomy. *Ophthalmology* 1992;99:805–812.
16. Fagerholm P, Fitzsimmons TD, Orndahl M, et al: Phototherapeutic keratectomy: Long term results in 166 eyes. *J Refract Corneal Surg* (Suppl) 1993; 9:76–81.
17. Campos M, Nielsen S, Szerenyi K, Garbus JJ, McDonnell PJ: Clinical follow-up of phototherapeutic keratectomy for treatment of corneal opacities. *Am J Ophthalmol* 1993;115:443–440.
18. Chamon W, Azar DT, Stark WJ, et al: Phototherapeutic keratectomy. *Ophthalmol Clin North Am* 1993;6:399–412.
19. Rapuano CJ, Laibson PR: Excimer laser phototherapeutic keratectomy. *CLAO J* 1993;19:235–240.
20. Wu WCS, Stark WJ, Green WR: Corneal wound healing after 193-nm excimer laser keratectomy. *Arch Ophthalmol* 1991;109:1426–1432.
21. Sanders D: Clinical evaluation of phototherapeutic keratectomy-VISX twenty/twenty excimer laser, written communication. In Salz JJ (ed): *Corneal Laser Surgery*. St Louis, CV Mosby, 1995, p 219.
22. Azar DT, Jain S, Woods K, et al: Phototherapeutic keratectomy: The VISX experience. In Salz JJ (ed): *Corneal Laser Surgery*. St Louis, CV Mosby, 1995, p 224.
23. Uozato H, Guyton DL, Waring GO: Centering corneal surgical procedures. *Am J Ophthalmol* 1987; 103:264–275.
24. Seiler T, Schmidt-Petersen H, Wollensak J: Complications after myopic photorefractive keratectomy, primarily with the Summit excimer laser. In Salz JJ (ed): *Corneal Laser Surgery*. St Louis, CV Mosby, 1995; pp 131–142.
25. Spadea L, Sabetti L, Balestrazzi E: Effect of centering excimer laser PRK on refractive results: A corneal topography study. *J Refract Corneal Surg* 1993; 9(Suppl):22–25.
26. Talamo JH, Wagoner MD, Lee SL: Management of ablation decentration following excimer photorefractive keratectomy. *Arch Ophthalmol* 1995 in press.
27. McDonnell JM, Garbus JJ, McDonnell PJ: Unsuccessful excimer laser phototherapeutic keratectomy. *Arch Ophthalmol* 1992;110:977–979.
28. Carones F, Brancato R, Venturi E, Morioc A: The corneal endothelium after myopic excimer laser photorefractive keratectomy. *Arch Ophthalmol* 1994;112:920–924.
29. Kermani O, Lubatschowski H: Struktur und dynamik photoakustischer shockwellen bei der 193 nm excimerlaserphotoablation der hornhaut. *Fortschr Ophthalmol* 1991;88:748–753.
30. Dehm EJ, Puliafito CA, Adler CM, Steinert RF: Corneal endothelial injury in rabbits following excimer laser ablation at 193 and 248 nm. *Arch Ophthalmol* 1986;104:1364–1368.
31. Fantes FE, Hanna KD, Waring GO III, et al: Wound

healing after excimer laser keratomileusis (photorefractive keratectomy) in monkeys. *Arch Ophthalmol* 1990;108:665–675.
32. Amano S, Shimizu K: Corneal endothelial changes after excimer laser photorefractive keratectomy. *Am J Ophthalmol* 1993;116:692–694.
33. Perez-Santonja JJ, Meza J, Moreno E, et al: Short-term corneal endothelial changes after photorefractive keratectomy. *J Refract Corneal Surg* 1994; 10 (Suppl):194–198.
34. Phillips AF, Szerenyi K, Campos M, et al: Arachidonic acid metabolites after excimer laser corneal surgery. *Arch Ophthalmol* 1993;111: 1273–1278.
35. Moreira H, McDonnell PJ, Fasano AP, et al: Treatment of experimental *Pseudomonas* keratitis with cyclo-oxygenase and lipoxygenase inhibitors. *Ophthalmology* 1991;98:1693–1697.
36. Arshinoff S, D'Addario, Sadler C, et al: Use of topical nonsteroidal anti-inflammatory drugs in excimer laser photorefractive keratectomy. *J Cataract Refract Surg* 1994;20:216–222.
37. Sher NA, Frantz JM, Talley A, et al: Topical diclofenac in the treatment of ocular pain after excimer photorefractive keratectomy. *J Refract Corneal Surg* 1993;9:425–436.
38. Ku EC, Lee M, Kothari HV, Scholer DW: Effect of diclofenac sodium on the arachidonic acid cascade. *Am J Med* 1986;80(Suppl 4B):18–23.
39. Szerenyi K, Sorken K, Garbus JJ, et al: Decrease in normal human corneal senstivity with topical diclofenac sodium. *Am J Ophthalmol* 1994;118: 312–315.
40. Szerenyi K, Wang XW, Lee M, McDonnell PJ: Topical diclofenac treatment prior to excimer laser photorefractive keratectomy in rabbits. *J Refract Corneal Surg* 1993;9:437–442.
41. Sher NA, Krueger RR, Teal P, et al: Role of topical corticosteroids and nonsteroidal anti-inflammatory drugs in the etilogy of stromal infiltrates after excimer photorefractive kerateactomy. *J Refract Corneal Surg* 1994;10:587–588.
42. Campos M, Abed HM, McDonnell PJ: Topical fluormetholone reduces stromal inflammation after phororefractive keratectomy. *Ophthalmic Surg* 1993; 24:654–657.
43. Campos M, Hertzog L, Garbus JJ, et al: Corneal sensitivity after photorefractive keratectomy. *Am J Ophthalmol* 1992;114:51–54.
44. Seiler T, Wollensak J: Myopic photorefractive keratectomy with the excimer laser: One year follow-up. *Ophthalmology* 1991;98:1156–1163.
45. Pavan-Langstone: Viral disease of the cornea and external eye. In Albert DM, Jakobiec FA (eds): *Principles and Practice of Ophthalmology—Clinical Practice*, Vol 1. Philadelphia, WB Saunders, 1994, chap 6.
46. McDonnell PJ, Moreira H, Clapham TN, et al: Photorefractive keratectomy for astigmatism. *Arch Ophthalmol* 1991;109:1370–1373.
47. Vrabec MP, Anderson JA, Rock ME, et al: Electron microscopic findings in a cornea with recurrence of herpes simplex keratitis after excimer laser phototherapeutic keratectomy. *CLAO J* 1994;20:41–44.
48. Pepose JS, Laycock KA, Miller JK, et al: Reactivation of latent herpes simplex virus by excimer laser photokeratectomy. *Am J Ophthalmol* 1992;114:45–50.
49. De La Maza MS: Immunoregulation, immune tolerance, autoimmunity, and immune "privilege." In Albert DM, Jakobiec FA (eds): *Principles and Practice of Ophthalmology. Basic Sciences*. Philadelphia, WB Saunders, 1994, chap 64.
50. Hersh PS, Jordan AJ, Mayers M: Corneal graft rejection episode after excimer laser photothérapeutic keratectomy. *Arch Ophthalmol* 1993;111:735–736.
51. Epstein RJ, Robin JB: Corneal graft rejection episode after excimer laser phototherapeutic keratectomy. *Arch Ophthalmol* 1994;112:157.

section four

PTK for PRK Complications

CHAPTER 11

Photorefractive Keratectomy (PRK) Outcomes and Complications

Tat Keong Chan, M. Farooq Ashraf, Dimitri T. Azar

INTRODUCTION

In addition to its therapeutic applications, the excimer laser has been proven effective in correcting refractive errors. The use of the 193-nm argon-fluoride excimer laser for the treatment of myopia has undergone a rapid evolution in the past decade. The original investigators of excimer technology were scientists in the computer industry who developed the excimer laser for the manufacture of semiconductor chips. In 1981, Toboada and Archibald noted the extreme sensitivity of the corneal epithelium to the excimer laser.[1] Four years later, Trokel and associates demonstrated that the laser could remove corneal tissue with extreme precision to alter the curvature of the cornea.[2] In 1987 the first photorefractive keratectomy (PRK) on human eyes was performed by McDonald and colleagues.[3]

Although the excimer laser was developed in the United States, it has been used to treat myopia extensively throughout the world, with excellent results. The Food and Drug Administration (FDA) has issued stringent guidelines for the clinical evaluation of this new technique in the United States. Two manufacturers, VISX (Santa Clara, CA) and Summit (Waltham, MA), have received FDA approval after completing phase III clinical trials.

Patient Selection

Patient selection had several similarities in the VISX and Summit U.S. clinical trials. Criteria for patient entry and exclusion in most other studies were based on the FDA clinical trials. Prospective patients were given extensive written information on surgical and nonsurgical options for the correction of myopia. Written material on the excimer laser, the investigational nature of the procedure, and all potential complications or side effects were reviewed. Patients in these studies were then counseled on the preoperative evaluations, the operative procedure, and the postoperative follow-up and evaluations prior to enrolling in the studies. Patients were also informed that early postoperative vision may be poor, early hyperopic shift may occur, and haze, glare, and overcorrection may develop.

FDA inclusion criteria for PRK phase III study patients stated that patients must be at least 18 years old, and have a best corrected visual acuity of 20/40 or better in both eyes and an astigmatic component of the manifest refraction of ≤1.00 D. Patients with contact lenses were required to remove them 2–3 weeks prior to baseline measurements. Stability of refraction was usually defined as a change of less than 0.50 D during the previous year. Additional criteria included a normal slit lamp examination and absence of prior corneal surgery or systemic disease that might impede corneal wound healing. As with any refractive procedure, the patients should have demonstrated proper motivation and understanding of the details and potential risks and benefits of the procedure.

Exclusion criteria in most studies included unstable refraction, keratoconus, uveitis, advanced dry eyes, severe belpharitis, lagophthalmos, marked corneal neovascularization, immunocompromised patients, and patients with connective tissue disorders.

Preoperative Evaluation

For phase III PRK patients, preoperative testing entailed measuring uncorrected and corrected visual acuity, manifest and cycloplegic refractions, keratometry, tonometry, corneal topography, slit lamp, and dilated funduscopic examinations. Pachymetry, glare testing, contrast sensitivity, visual fields testing, and endothelial cell counts were performed in some study centers.

Surgical Technique

Preoperative topical medications usually included a topical anesthetic and topical antibiotics. Some surgeons used topical pilocarpine. Oral pain medications and sedatives such as diazepam were also given in some studies. Prior to surgery, a careful calibration of the laser was performed. Test ablations on polymethyl methacrylate (PMMA) blocks were performed and checked using a lensometer to ensure adequate depth and consistency of the desired ablation. The desired numeric parameters for myopic correction, optical zone size, and ablation rate setting were entered in the computer software. Fixation was predominantly achieved by patient self-fixation. A circular optical zone mark was used to mark the treatment area. Epithelial removal was performed mainly by mechanical debridement with a blade, a spatula, or methylcellulose sponges.

The argon-fluoride laser with spectral emission of 193 nm has a pulse frequency at 5 or 10 Hz. The pulse energy results in a radiant energy exposure of 100–180 mJ/cm^2. The number of pulses for the desired correction was computed based on algorithms supplied by the laser manufacturers. The ablation zone varied from 3.5 to 6.2 mm in diameter. Most studies used single-zone treatments; however, some studies did use multizone treatments.

Postoperative Management

At the completion of ablation, the lid speculum was removed and topical ophthalmic drops instilled into the operative eye. These drops usually included an antibiotic, a cycloplegic, steroids, and/or nonsteroidal anti-inflammatory agents. Oral analgesics were also prescribed for postoperative pain. A pressure patch or a bandage contact lens was placed, and patients were seen daily until complete reepithelialization occurred. Corticosteroids, usually fluorometholone or dexamethasone, were continued in a tapering fashion for 3–6 months. At some study centers, patients were followed at 1, 3, 6, 9, and 12 months postoperatively. Postoperative measurements included visual acuity, refraction, keratometry, tonometry, and detailed slit lamp examination. Postoperative subepithelial corneal haze was graded subjectively at each center for each postoperative visit.

RESULTS OF CLINICAL STUDIES

Animal studies using the 193-nm argon-fluoride excimer laser have provided the scientific basis for subsequent human clinical trials in the treatment of myopia.[4-8] Since the time of the excimer laser studies on blind and partially sighted eyes, many clinical trials have been performed in sighted eyes.[9-13] The following is a summary of the results of some of the major clinical studies done to date using the Summit excimer laser system and the VISX excimer laser system (Table 11.1). Other studies have been arbitrarily divided into three categories: low-myopia studies (≤6 D; Table 11.2), high-myopia studies (>6 D; Table 11.3), and combined low- and high-myopia studies (1–25 D; Table 11.4). Since a description of the results of every clinical study performed to date is beyond the scope of this book, the following list is by no means exhaustive.

REFRACTIVE OUTCOME

Low-Myopia Studies

The first clinical trials to be performed using lasers from either laser company were in eyes with low myopia, i.e., 6 D of myopia or less. One of the earliest sighted eye studies was reported by McDonald and coworkers in 1991.[11] They performed PRK on 19 sighted eyes as part of a series which also included 10 partially sighted eyes. Seventeen of the sighted myopic eyes had at least 12 months of follow-up. Attempted corrections in the low-myopia group ranged from −2.25 to −5.00 D, although the range of attempted correction in the overall study extended to −8.00 D. The VISX 20/20 excimer laser system was used to perform PRK, with an

Table 11.1. Results of Summit and VISX U.S. PRK Clinical Studies

	No. of Eyes	Mean F/U (months)	Single/ Multi Zone	OZ (mm)	Attempted Correction (D)	±1 D of Emmetropia	UCVA 20/40	Loss of BCVA >2 lines	Increased IOP
Summit	544	12	Single	6.00	−1 to −6	76%	91%	1%	
VISX	691	12	Single	6.00	−1 to −6	79%	86%	1%	3% >5 mm Hg
	691	24	Single	6.00	−1 to −6	79%	85%	1%	0.2% >10 mm Hg

F/U, follow-up; D, diopter; UCVA, uncorrected visual acuity; BCVA, best corrected visual acuity; IOP, intraocular pressure.

ablation zone of 5.00 mm in the low-myopia group. Results showed early overcorrection within the first 2 months followed by regression; at 12 months, 57% of the patients were within ±1 D of the attempted correction.[11]

Gartry et al.[14] reported on the results of their treatments with the Summit Technology UV200 excimer laser system of 120 eyes with ≥18 months follow-up. Preoperative refractions ranged from −1.50 to −17.50 D. In the low-myopia group, there were 20 eyes in each subgroup treated from −2.00 D through −6.00 D in whole-diopter steps. Within each subgroup, the same attempted correction of myopia was performed. A rather small ablation zone of 4.00 mm was used. The authors found that the predictability of PRK diminished as larger corrections were attempted (Table 11.2). One important reason for the greater regression seen in the more myopic subgroup, the authors postulated, was the more marked tissue healing responses resulting from deeper ablations.[14]

Brancato and the Italian Study Group reported on a large multicenter study using the Summit laser system.[15] The attempted correction ranged from −0.80 to −25.00 D, and 330 eyes had at least 12 months of follow-up. A single ablation zone ranging from 3.5 to 5.00 mm was used in treatments of −6.00 D or less. In the low-myopia subgroup, in which treatment was between −0.80 and −6.00 D, the mean spherical equivalent error was −0.52 ± 1.04 D at 12 months. In this group, 71.2% were within ± 1 D of the attempted correction.

In Asia, Kim and coauthors[16] reported the 1-year results of a study in 202 eyes undergoing PRK using the Summit laser with a maximum attempted treatment of −6.00 D. The patients were divided into two groups: group 1, in which the preoperative refraction was −2.00 to −7.00 D, and group 2, in which the preoperative refraction was −7.25 to −13.50 D. In group 1, the attempted correction was based on the preoperative myopic state. In group 2, single-step −6.00 D PRK was performed. Of the eyes in group 1, 91.4% were within ± 1 D of the intended result, whereas the corresponding figure was only 51.7% in group 2 eyes. THe authors concluded that PRK was an extremely effective method for correcting myopia in eyes below −7.00 D but was less predictable in eyes with higher levels of myopia.

In Maguen et al's series of 122 eyes treated with the VISX excimer laser with a minimum of 12 months' follow-up, 79% were within ± 1 D of the intended correction.[17] This figure increased to 86% and 90% at 24 and 36 months, respectively. The preoperative spherical equivalent ranged from −1.00 to −7.75 D, although the maximum amount of correction was also −6.00 D. Emmetropia was therefore not the aim in correcting some of the more myopic eyes. Scattergrams of the achieved correction versus intended correction at different times periods show that there was less scatter in the results seen in the −1.00 to −3.50 D group compared with the group with preoperative myopia above −3.50 D.

In a study comparing the performance of the Summit Excimer 200 LA and the VISX 20/20 excimer lasers, Hamberg-Nystrom et al. studied 20 eyes with low myopia in each group treated using a 5.00-mm ablation zone.[18] Emmetropia was attempted in all eyes. The results at 12 months showed a median refraction of plano in the Summit group compared to −0.50 D in the VISX group.

In an attempt to study the effects of ablation zone diameter on the outcome of PRK, O'Brart and coworkers[19] examined a series of 33 patients who had bilateral PRK using different ablation zones.

Table 11.2. Low-Myopia Studies

Author/ Laser/Year	Low Myopes/ (No. of Eyes in Main Study)	Mean Follow-up (mo)	Attempted Correction (D)	Single/ Multi Zone (mm)	OZ (mm)	UCVA 20/40 (%)	UCVA 20/20 (%)
McDonald (USA) VISX/1991[11]	7(18)	12	−2.25 to −5.00	Single	5.00	86	
Gartry (UK) Summit/1992[14]	120 (120) 20 20 20 20 20	18	−2.00 to −7.00 −2.00 −3.00 −4.00 −5.00 −6.00	Single	4.00	See below 90 78 59 63 63	See below 55 56 47 50 25
Brancato (Italy) Summit/1993[15]	146 (330)	12	−0.8 to −6.00	Single	3.50 to 5.00	Mean UCVA 20/32 at 12 mo	
Kim (Korea) Summit/1993[16]	135 (202) Preop myopia: −2.00 D to −7.00 D (Gp 1)	12	−2.00 to −6.00	Single	5.00	98.5	
	67 (202) Preop myopia: −7.25 D to −13.50 D (Gp 2)	12	−6.00	Single	5.00	62.7	
Maguen (USA) VISX/1994[17]	122 (122)	12	−1.00 to −6.00	Single	? 6.00	89	37
	48 (48) 9 (9)	24 36				92 90	48 80
Hamberg-Nystrom (Sweden) Summit/1994[18]	20 (20)	12	−2.00 to −5.00	Single	5.00	100	80
Hamberg-Nystrom (Sweden) VISX/1994[18]	20 (20)	12	−2.00 to −5.00	Single	5.00	85	35
O'Brart (UK) Summit/1994[19]	33 (33) first eyes	24	−2.00 to −6.00	Single	4.00	Not studied	Not studied
	33 (33) second eyes	12	−2.00 to −6.00	Single	5.00	Not studied	Not studied

± 1 D (%)	Corneal Haze	Topical Steroids	Increased IOP (%)	Loss BCVA 2 Lines (%)	Loss BCVA >2 Lines (%)	Remarks/Comments
57	0% >grade 1+ haze	Yes	0	0	0	
See below 95 70 40 50 40	Maximum at 6 mo	Yes	12 (overall study)	3	0	Halo effect in 78%, major halo problem in 10%. Significant decentration in 1% due to patient movement. In the −2 to −5 D groups, refraction stable after 3 mo. In the −7 D group, 2 D of regression from 3 to 6 mo.
71.2	Haze level proportional to attempted correction	Yes	25.3	2.4	0	9.4% had late epithelial healing (>4 days); 10.6% had decentration of ablation zone >1 mm.
91.4	0% with grade 2 haze or worse	Yes	14.1	8.1 (2 lines or more)		Overall, glare or haloes seen in 10%. Myopic regression seen in 4% of the overall study group.
51.7	3% with grade 2 haze or worse	Yes	23.9	17.9 (2 lines or more)		Same as above.
79 (12 mo) 86 (24 mo) 90 (36 mo)	Haze maximum from 1–6 mo. Most common grading is 1.5.	Yes	10.8 (at any time)	4 (12 mths)	0 (12 mo)	One-third of eyes received N_2 flow and two-thirds did not. 5% of all eyes developed central islands; all such eyes did not receive N_2 flow. In 1.25% of eyes, the islands persisted beyond 6 mo. Two eyes developed myopic regression and 1 eye was retreated.
Not studied	Not studied	Yes	Maximum 18 mm Hg	Not studied	Not studied	Median spherical equivalent at 12 mo was 0.00 D. This result was statistically different from that of the VISX laser (−0.5 D). No statistical difference in intraocular pressures, contrast sensitivity, or centration.
Not studied	Not studied	Yes	Maximum 21 mm Hg	Not studied	Not studied	Median spherical equivalent at 12 mo was −0.50 D. Difference from that of Summit laser (0.00 D) statistically significant.
0 > + 1.00 D	Maximum at 3 mo	Yes (19 eyes) No (14 eyes)	Not reported	0	0	Magnitude of haloes, as measured with a computer program, statistically less in eyes treated with 5.00-mm ablation zone. Mean postoperative refractive change significantly greater for 5-mm ablation zone ($p < 0.01$). No significant difference between the two groups in terms of anterior stromal haze.
15 > + 1.00 D	Maximum at 3 mo	Yes (19 eyes) No (14 eyes)	Not reported			
				0	0	

Table 11.3. High-Myopia Studies

Author/ Laser/Year	High Myopes/ (Overall No. of Eyes)	Mean Follow-up (mo)	Attempted Correction (D)	Single/ Multi Zone (mm)	OZ (mm)	UCVA 20/40 (%)	UCVA 20/20 (%)
McDonald (USA) VISX/1991[11]	11 (18)	12	−5.01 to −8.00	Single	5.00 (−6.00 D	18 or less) 4.50 (−6.01 to −7.00D) 4.25 (−7.01 to −8.00 D)	
Brancato (Italy) Summit/1993[15]	145 (330)	12	−6.10 to −9.90	Single	3.50 to 5.00	Mean UCVA at 12 mo was 20/63	
	39 (330)	12	−10.00 to −25.00	Double	First zone 3.50 to 4.80 Second zone 3.80 to 5.00	Mean UCVA at 12 mo was 20/63	
Heitzman (USA) VISX/1993[20]	23 (23)	7.5	−8.00 to −19.50	Multizone	4.00, 5.00, 6.00	52	
Buratto (Italy) (Summit/1993[21]	40 (40)	24	−6.00 to −10.00	Single	4.30 or 4.50	Not reported	Not reported
Sher (USA) VISX/1994[22]	47 (47)	6 to 12	−8.00 to −15.25 (mean −11.2)	Single	5.50 to 6.20	60	Not reported
Talamo (USA) VISX/1995[23]	46	6	−6.00 to −8.00 mean −6.98)	Single (67%) Double (33%)	6.00 (single) 6.00 and 5.00 (double)	74	17

± 1 D (%)	Corneal Haze	Topical Steroids	Increased IOP (%)	Loss BCVA 2 Lines (%)	Loss BCVA >2 Lines (%)	Remarks/Comments
18	5.8% had > grade	Yes 1+ haze	0	0	0	One eye lost 3 lines BCVA due to irregular astigmatism; 2 lines regained by hard contact lens refraction. Another eye lost 2 lines of BCVA due to irregular astigmatism from a decentered ablation; both lines regained with hard contact lens refraction.
34.5 28.2	Positive correlation between corneal haze and attempted correction	Yes	25.3 (overall study)	2.4 (overall study)	0	In the overall study, 9.4% had late reepithelialization (>4 days) and 10.6% had decentration of the ablation zone >1 mm.
39	Peak at 3 mo. 8.7% grade 2+ haze	Yes	0	15	0	Loss of BCVA due to irregular astigmatism and corneal haze. One eye had decentration of ablation zone of 2.4 mm. 12% developed central islands. Two eyes had myopic regression and were retreated. One eye underwent radial keratotomy for regression.
35	grade 1+ in 77.5% grade 2+ in 7.5%	Yes	7.5	40		Mean SE was −8.12 D preoperatively and −2.31 D at 24 mo. Haloes in 5%, glare in 10%, irregular astigmatism in 2.5%. 65% had myopic regression exceeding −1.00 D. UCVA was 20/100 in 60%.
58 (excluding retreatments) 47 (all cases)	Peak at 3 mo. 8.7% grade 2+	Yes	0	15	0	Irregular astigmatism and corneal haze were causes of loss of BCVA. One eye had decentration of ablation zone of 2.4 mm. 12% developed central islands.
67	11% grade 1+ or more	Yes	0	2 (irregular astigmatism		Single-zone ablations had better refractive and visual acuity outcomes than double-zone ablations. Mean UCVA was 20/30 and 20/50 for single-zone and double-zone ablations, respectively. Also, 76% and 36% were within 1 D of emmetropia, respectively.

Table 11.4. Combined Low- and High-Myopia Studies

Author/Laser/Year	No. of Eyes in Study	Mean Follow-up (mo)	Attempted Correction (D)	Single/Multi Zone	OZ (mm)	UCVA 20/40 (%)	UCVA 20/20 (%)
Seiler (Germany) Summit/1991[24]	26	12	−1.4 to −9.25 (mean −4.5)	Single	3.5	96 (except 3 eyes intentionally undercorrected)	47.8 (except three eyes intentionally undercorrected)
Tengroth (Sweden) Summit/1993[25]	420	12	−1.25 to −7.50 (mean −4.03)	Single	4.50 (less than −5.5 D) 4.30 (−5.5 D or more)	91	Not reported
Ficker (UK) Summit/1993[26]	61	12	−1.00 to −10.00	Single	5.00 (−8.00 D or less) 4.50 (more than −8.00 D)	81	Not reported
Epstein (Sweden) Summit/1994[27]	495	24	−1.25 to −7.50 (mean 4.05)	Single	4.50 (less than 5.50 D) 4.30 (5.50 to 7.50 D)	91	Not reported

The Summit excimer laser was used to perform identical dioptric corrections in both eyes, with first eyes being treated with 4.00-mm ablation zones and second eyes with 5.00-mm ablation zones. The authors found that mean changes in postoperative refraction were significantly greater in eyes treated with 5.00-mm ablation zones. No eyes treated with 4.00-mm ablation zones were overcorrected, whereas 15% of eyes treated with 5.00-mm ablation zones had refractions above + 1.00 D 12 months after PRK. The authors asserted that this trend may reflect a less aggressive epithelial wound-healing tendency associated with larger-beam diameters.

High-Myopia Studies

In McDonald's study mentioned above, there were 11 highly myopic sighted eyes with attempted corrections ranging from −5.01 to −8.00 D. Ablation zones varied in diameter from 4.25 to 5.00 mm. In this high-myopia subgroup, only 2 of 11 eyes (18%) achieved a final refraction within ± 1 D of the attempted correction at 12 months, a much lower percentage than in eyes treated for 5 D of myopia or less. Predictability in the high-myopia group was poor, with 8 of 11 eyes (73%) having achieved spherical equivalents differing by more than ± 3D from the intended results.[11]

In Brancato and the Italian Study Group's multicenter study, 34.5% of eyes with attempted corrections ranging from −6.10 to −9.90 D were within ± 1 D of the intended result at 12 months.[15] (A single ablation zone ranging from 3.50 to 5.00 mm was used in this subgroup.) In contrast, only 28.2% of eyes with attempted corrections ranging from

± 1 D (%)	Corneal Haze	Topical Steroids	Increased IOP (%)	Loss BCVA 2 Lines (%)	Loss BCVA >2 Lines (%)	Remarks/ Comments
92	All eyes except one had grade 1+ or less haze	Yes	3.1	0	0	One eye developed myopic regression with grade 1+ haze and was retreated. Six patients had persistent haloes.
86	Mean haze at 12 mo was 0.77+. 3 or 4+; haze always correlated to regression. Haze reversed with topical steroids	In a substudy: Group 1: treated for 3 mo Group 2: treated for 5 weeks	13	Not reported	Not reported	Group 1 eyes regressed significantly less than Group 2 eyes ($p < 0.01$). Glare and haloes not a significant problem.
81	Mean grading at 12 mo): 1+	Yes	Not reported	0	0	Regression to myopia more common in eyes treated for more than −4.00 D. Glare not a significant problem.
87.5	Grade 2+:in 3% Grade 3+:in 1%	Yes	13	0	0	Subgroup analysis showed that eyes with low myopia (less than −3.9 D) had significantly better refractive outcomes than eyes with higher myopia.

−10.00 to −25.00 D were within the same level of accuracy. In the latter subgroup of highly myopic eyes, an ablation zone of 3.50 to 4.80 mm was used first, followed by a second ablation zone ranging from 3.80 to 5.00 mm. More regression was seen in eyes with attempted corrections above −6.00 D compared to eyes with attempted corrections below −6.00 D.

Heitzmann et al. studied 23 highly myopic eyes which underwent multizone (4.00, 5.00, 6.00 mm) PRK using the VISX laser system.[20] The preoperative spherical equivalents ranged from −8.00 to −19.50 D, and the mean follow-up was 7.5 months. At the last postoperative examination, the mean spherical equivalent was −1.09 ± 2.08 D, including results from two retreatment cases, and only 39% of eyes were within ± 1 D of emmetropia. In terms of stability, 7 of 18 eyes (39%) with at least 6 months of follow-up exhibited significant regression, with a refractive change of 1.75 D or more. Of these, two eyes were retreated and one eye had radial keratotomy for residual myopia. When comparing their results with those of other reported studies, the authors contended that a multizone approach to reduce ablation depth in PRK for high myopia did not appear to decrease myopic regression.

In a retrospective study of 40 consecutive eyes that underwent PRK for myopia ranging from −6.00 to −10.00 D using the Summit laser, Buratto and Ferrari reported that only 35% of eyes were within ± 1 D of the desired correction at 24 months.[21] A single-zone technique and a small ablation zone of either 4.30 or 4.50 mm were used. The mean postoperative spherical equivalent was −2.31

± 1.25 D. In 65% of the eyes there was myopic regression after 6 months of follow-up. The authors thus observed that PRK had poor predictability as a result of significant regression of the refractive effect in their series. They also suggested that keratomileusis in situ combined with the excimer laser might be a better form of treatment for highly myopic eyes.

Using the VISX laser and larger ablation zones of mainly 6.00 to 6.20 mm and a single-zone technique, Sher and coauthors evaluated 47 eyes which underwent PRK as part of the FDA's phase II B study.[22] Preoperative myopia ranged from −8.00 to −15.25 D (mean, −11.20 D); 23% of patients were retreated 6 to 16 months later for undercorrection or regression. After retreatment, 47% of all eyes were within 1 D of the attempted correction. At 12 months, 58% of the eyes that had not been retreated achieved correction within 1 D of the attempted correction. The authors also noted that there was a greater incidence of undercorrection in eyes treated with a beam diameter of 5.5 to 5.8 mm compared to eyes treated with a larger diameter of 6.00 to 6.20 mm. The refractive change seen with a larger ablation zone was similar to the observation by O'Brart mentioned previously.[19]

More recently, Talamo and the VISX Moderate Myopia Study Group reported the results of a multicenter PRK study for myopia of 6.00 to 8.00 D, with a mean of −6.98 D.[23] Forty-six eyes achieved at least 6 months of follow-up, and 67% were within ± 1 D of the intended correction. This percentage was higher than the corresponding figures in all the other high-myopia studies mentioned previously because the attempted correction of myopia in this case was lower, being in the moderate rather than the high-myopia group. Having used large-diameter ablation zones in this study (6.00 mm for single-zone ablations and 6.00 and 5.00 mm for double-zone ablations), the authors concluded that PRK for moderate myopia using large ablation zones was more predictable than when smaller ablation zones were used.

Combined Low- and High-Myopia Studies

In one of the earliest studies of PRK using the Summit laser, Seiler and Wollensak reported on 26 eyes treated for myopia ranging from 1.40 to 7.25 D.[24] Emmetropia was the aim of treatment in all except three highly myopic eyes. A small ablation zone of 3.5 mm was used. Overall, at 1 year, 92% of the eyes were within ± 1 D of the intended result. Predictability was much higher in the low-myopia group (preoperative refraction below −5.00 D) than in the group whose preoperative refraction was above −5.00 D (100% versus 81.8%).

A larger study of 420 eyes with preoperative refraction ranging from −1.25 to −7.50 D treated with the Summit laser was reported by Tengroth et al.[25] Minimum follow-up was 12 months. A 4.50-mm ablation zone was used in eyes with up to 5.40 D of preoperative myopia, and a 4.30-mm ablation zone was used in eyes with preoperative myopia of 5.50 D or more. Overall, 86% of the eyes were within ± 1 D of the intended correction, a figure comparable to that in Seiler's earlier study. Eyes with preoperative myopia of up to 4.90 D had significantly better refractive outcomes than eyes with preoperative myopia of 5.00 to 7.50 D.[25]

In a prospective study of PRK for myopic eyes between −1.00 and −10.00 D using the Summit laser, Ficker and coauthors followed 61 eyes for 12 months.[26] The ablation zone was 5.00 mm for myopia not exceeding 8.00 D and 4.5 mm for myopia above 8.00 D. The authors found that 81% of the eyes were within ± 1 D of emmetropia, and myopic regression was more common among moderate and high myopes (above −4.00 D).

Epstein et al. reported on a series of 495 eyes undergoing PRK using the Summit laser with follow-up of 24 months.[27] Preoperative refractions ranged from −1.25 to −7.50 D (mean, −4.05 D), and ablation zones of 4.30 and 4.50 mm were used. Mean refraction at 24 months was −0.27 ± 0.74 D, which was significantly different from that at 12 months (0.01 ± 0.78 D). In this series, 87.5% of the eyes were within ± 1 D of emmetropia, and subgroup analysis showed that eyes with low to moderate myopia (less than −4.00 D) had significantly better refractive outcomes than those with higher myopia. This observation is essentially similar to those found in many of the previously mentioned studies.[11,14,15,24–26]

UNCORRECTED VISUAL ACUITY

Low-Myopia Studies

The uncorrected visual acuity (UCVA) after PRK is dependent on the final refractive outcome and varies with the amount of correction attempted. In Table 11.2, the percentage of eyes with UCVA of 20/40 or better varies from 63% in two subgroups in Gartry's series to 100% in Hamberg-Nystrom's Summit series.[18] It can be seen that for low myopes in most of the studies mentioned, the percentage of eyes with UCVA of 20/40 or better is very high. Most authors concur that PRK is a very effective modality of treatment for low myopia.

High-Myopia Studies

The percentage of eyes with UCVA of 20/40 or better in this group was lower than that in the low-myopia group. From Table 11.3, this ranged from only 18% in MacDonald's early series of 11 eyes to 60% in Sher's series of 47 eyes.[11,22] In Talamo's series of 46 eyes with a mean attempted correction lower than that in Sher's study (−6.98 D versus −11.20 D), 74% had a UCVA of 20/40 or better.[23]

Combined Low- and High-Myopia Studies

The percentage of eyes achieving UCVA of 20/40 or better ranged from 81% in Ficker's study to 96% in Seiler's study.[24,26] In Seiler's study, this percentage did not include eyes that were intentionally undercorrected. In this same study, visual acuity from glare decreased from 20/27 preoperatively to 20/31 after 12 months, a result attributable to the presence of corneal haze and a small ablation zone diameter (3.50 mm).

BEST SPECTACLE CORRECTED VISUAL ACUITY

Work by Zadnick et al. has shown that the loss of two lines of best spectacle corrected visual acuity (BCVA) is considered statistically significant.[28] Loss of two lines of BCVA ranged from 0% (such as in studies by Seiler, Ficker, and Epstein) to as high as 40% in a high-myopia study by Buratto[21,24,26,27] (Tables 13.2, 13.3, and 13.4) The most commonly cited causes of this problem were corneal haze, decentration of the ablation zone, and resultant irregular astigmatism. Brancato's study, however, showed no statistical difference between corneal haze and BCVA, suggesting that corneal haze in that study did not seem to interfere with long-term visual acuity.[15] The loss of BCVA due to irregular astigmatism was usually regained with the wearing of a rigid gas-permeable contact lens.[11]

CORNEAL CLARITY AND THE ROLE OF POSTOPERATIVE TOPICAL STEROIDS

Corneal haze was reported to be maximum during the first 6 months after PRK, decreasing to negligible levels by about 18 months.[14,17–19,29] Its severity was found to be proportional to the attempted correction of myopia.[15,16,29] It was not a serious problem in the low myopes, with the incidence of grade 2+ haze or more being 0% at 12 months in both Gartry's study and McDonald's low-myopia subgroup of eyes.[11,14] In Kim's study, the incidence of grade 2+ haze or greater was 0% in the low-myopia subgroup (attempted correction, −2.00 to −6.00 D) and 3% in the high-myopia subgroup (preoperative refraction, −7.25 to −13.50 D) treated with a maximum of −6.00 D.[16] In a study to assess the cause of subepithelial corneal haze over an 18-month period after PRK, Caubet reported that the overall frequency of clinically significant haze was 11.5%, and that haze was statistically higher in men and in ablations less than 4.50 mm in diameter.[29] Caubet did not find a statistical difference in the amount of haze in the different age groups.

The effect of topical steroids on refraction and corneal haze remains controversial. On the one hand, studies have shown that severe haze was associated with regression of the myopic effect and was in most cases reversible with the use of topical steroids.[14,25,30–32] Tengroth and coauthors found that eyes treated with topical steroids for 3 months regressed significantly less than eyes treated with topical steroids for only 5 weeks.[25] In the same study, topical steroids were also found to decrease the amount of postoperative corneal haze substantially. In a paper from the same institute by Fagerholm et al., the refractive outcome in 100 eyes that did not receive steroids following PRK was compared with that in 100 eyes that treated with topical dexamethasone for 3 months postoperatively.[33] PRK was performed with the Summit laser using small ablation zones of 4.3 and 4.5 mm. The two groups were matched for age, sex, and preoperative refractions. Results showed that at the 6-month follow-up, 86% of the untreated eyes had regressed to myopia of at least 0.50 D, whereas only 23% of the steroid-treated group had regressed to an equivalent amount of myopia. Subepithelial haze was also found to be more severe in the untreated eyes than in the steroid-treated eyes following PRK. The authors concluded that in their series of patients treated with relatively small ablation diameters, eyes that received topical dexamethasone had less corneal haze and were less likely to develop myopic regression.[33]

On the other hand, Gartry and coworkers found little justification in prescribing topical steroids after PRK.[30] In a prospective, randomized, double-masked clinical trial, they studied the effect of topical dexamethasone 0.1% on refraction and anterior stromal haze following −3.00 D and −6.00 D PRK treatments in 113 eyes using a 4.00-mm ablation zone. The authors found that initially the reduction of myopia was significantly greater in the steroid-treated group than in the placebo group. This difference became statistically insignificant on discontinuation of steroid therapy at 3 months. There was also no statistically significant difference in corneal haze between the steroid-treated group and the placebo group at any stage. There was, however, a significant statistical correlation between anterior stromal haze and the depth of induced ablation and between haze and myopic regression. The authors concluded that topical steroids should not be used following PRK. Similarly, in the study reported by Ficker et al., topical steroids did not seem to affect the postoperative refractive outcome.[26]

There was no statistically significant difference between corneal haze and the size of the ablation zone, as reported in O'Brart's study (4.00-mm versus 5.00-mm ablation zones).[19] Tavola et al. compared single- with double-zone PRKs in 166 eyes with attempted corrections between −6.50 D and −10.00 D using the Summit laser.[34] Cornea haze was found to be significantly more severe in the single-zone than in the double-zone ablations at 6 months. In another randomized study by Scialdone et al. with attempted corrections between −6.00 D and −9.00 D in 24 eyes, no significant difference in corneal haze was found between single- and double-zone ablations.[35] Scialdone's study, however, had a longer period of follow-up of up to 1 year, although the sample size was smaller.

Management of corneal haze usually includes waiting for the opacity to resolve spontaneously. In cases where the corneal haze is dense or where there is subepithelial fibrosis, topical steroids may be increased in potency and/or frequency or restarted if they have been stopped. In cases where there is myopic regression associated with persistent haze unresponsive to topical steroids, PRK retreatment is useful.[36,37] Seiler et al. reported on a series of 30 eyes retreated with the Summit laser for persistent corneal scarring and/or undercorrection.[36] The retreatments were performed at least 6 months after the first treatment. After a mean follow-up of 7.8 months, only one eye had developed mild corneal scarring following retreatment. Also, 63% of eyes were within ± 1 D of emmetropia 6 months after retreatment.

OCULAR HYPERTENSION

One of the major side effects of PRK is ocular hypertension related to the use of topical steroids. Its incidence ranged from 0.27% to 25.3% in Brancato's study.[15,38] The raised intraocular pressure was usually transient, and subsided after topical steroid therapy was discontinued and/or anti-glaucoma medical therapy commenced. However, visual field defects and glaucomatous nerve cupping occasionally occurred, as was reported by Kim et al.[38] Two eyes in this study underwent trabeculectomy for steroid-induced glaucoma 10 and 11 months following PRK.

GLARE AND HALOES

Glare and haloes are subjective visual complaints attributed by many to the scattering of light by opacities and surface irregularities of the postoperative cornea. These conditions occur more commonly at night, when the pupil enlarges, and may also manifest as a "starburst" phenomenon around bright sources of light. As some patients with essentially clear corneas following PRK still complain of disturbances in night vision, some authors consider glare and haloes an optical-refractive complication rather than a complication resulting entirely from a reduction in corneal transparency.[39] In a retrospective study of 15 eyes with clear cornea following PRK with the Summit laser, Seiler et al. found that complaints of glare and haloes correlated with the spherical aberration of the centrally flattened cornea.[40] It was the spherical aberration of the cornea which resulted in a degradation of the retinal image, causing the patient to complain of glare and haloes. Various studies have reported the incidence of post-PRK haloes ranging from 5% to 78%, depending on the time interval between laser treatment and the time the complaint was reported.[14,16,21]

Lohmann and coworkers described an objective method of measuring haloes using a computerized technique.[41] They used this method to measure haloes in four groups: spectacle lens wearers, soft contact lens wearers, hard contact lens wearers, and post-PRK patients with a follow-up period of 18 months. The study found that there were only marginal differences in halo size between the spectacle, soft contact lens, and post-PRK subgroups. All three groups were far superior to the hard contact lens subgroup, who experienced the largest haloes. Using the same objective method for measuring haloes, O'Brart et al. reported that the incidence of slight disturbances in night vision in 85 patients 6 months after PRK was 45%.[42] Eleven percent of all patients experienced significant problems with night vision. In assessing 43 patients who had undergone bilateral PRK, the first eye with a 4.00-mm ablation zone and the second eye with a 5.00-mm ablation zone, the authors found that 56% had smaller haloes and better night vision in the eye treated with a 5.00-mm zone. In comparison, 33% of patients experienced no difference between the two eyes and 12% reported that night vision was worse in the eye with the 5.00-mm zone. The results were statistically significant in favor of the 5.00-mm ablation zone. Another study reported later by the same authors also supported the conclusion favoring a larger ablation zone.[19]

The management of post-PRK haloes depends on the cause. If the cause of haloes is corneal haze and the symptomatic eye has residual myopia, reinstituting topical steroid therapy and even PRK retreatment can be considered. As haloes are known to occur more frequently in cases where the ablation zone is small, such as 5.00 mm or less, the ablation zone may be increased during PRK retreatment. A blended peripheral transition zone can be subsequently employed to minimize the depth of laser ablation.[42] If the eye is emmetropic or hyperopic, a trial of negative lens overcorrection or weak miotics at night may prove beneficial.[42]

DECENTRATION

As in all other keratorefractive procedures, centration of the cornea is of paramount importance in PRK. Most surgeons now routinely center the optical zone of the cornea around the center of the entrance pupil rather than around the visual axis of the eye.[43,44] Decentration of the ablation zone in PRK may result from patient movement during the surgical procedure or from the surgeon's failure to mark the optical center of the cornea accurately. Large decentrations (>1 mm) have been reported to cause significant postoperative irregular astigmatism, leading to reduced UCVA and/or BCVA.[20,45–48]

In a high-myopia PRK study, Sher et al. reported that decentration of 1.00 mm or more was seen in 32% of eyes, whereas decentration of 1.50 mm or more was present in 10% of eyes 6 months postoperatively.[22] Only one patient lost one line of BCVA, which resulted from a decentration of 2.40 mm. This patient had some complaints about night glare but, surprisingly, had few other visual complaints.

Cavanaugh et al. presented a technique for centration of the Summit Technology Excimer UV200 excimer laser and postoperative cornea topographic data on 110 patients from phase IIB and III clinical trials.[48] With the pupil constricted with topical 1% pilocarpine, the patient was instructed to maintain fixation on the green light in the laser apparatus. Centration was achieved by focusing the helium-neon beam at the corneal plane over the center of the entrance pupil. The two projection spots of the crossed helium-neon beams would then fall on the 3 and 9 o'clock positions on the iris at the pupillary margins. A 6.00-mm ablation zone was used for the PRK. The authors found that 92.73% of eyes were centered within 1.00 mm, while 57.27% were centered within 0.50 mm. Mean UCVA was 20/20 for decentrations of up to 1.00 mm. For decentrations exceeding 1.00 mm, the mean UCVA decreased to 20/30; the BCVA was also compromised. One drawback of the study was the fact that the corneal vertex was taken as the reference point for centration rather than the center of the pupil.

Klyce and Smolek analyzed the cornea topographic data on 17 eyes which had undergone PRK with the VISX 20/20 excimer laser at the LSU Eye Center in Los Angeles as part of the phase IIA trial.[49] The ablation decentration averaged 0.88 ± 0.11 mm from the center of the pupil. However, when the procedure to locate the cornea overlying the center of the pupil was refined in the phase IIB trial, decentration of the ablation zone was greatly minimized.

Lin and coauthors reported on the cornea topographic results in 97 consecutive eyes 1 month following PRK using the Topographic Modeling System (Computed Anatomy, Inc., NY).[50] They found that 85% of the eyes were within 0.50 mm of the pupillary center, 13% were between 0.50 and 1.00 mm, and 2% were more than 1.00 mm away from the pupillary center. The greatest amount of decentration was 1.50 mm, and the mean decentration was 0.36 mm. The surface regularity index (SRI), an index of the cornea's optical quality, was significantly less at 6 months that at 1 month after PRK. There was no correlation between the SRI and the amount of decentration in this study.[50]

Mild decentration of the ablation zone is usually innocuous and can be left alone. Large decentrations, however, are problematic and difficult to correct. A rigid gas-permeable contact lens may reduce the loss of visual acuity caused by irregular astigmatism. If there is significant undercorrection in addition to decentration, retreatment using a larger ablation zone to include the decentered area may be considered.[51]

CENTRAL ISLANDS

A central island is a postexcimer laser ablation topographic finding characterized by an area of corneal steepening within the ablative zone. It may be an important cause of irregular astigmatism leading to a loss of BCVA during the early postoperative period. Its definition varies among various authors.[22,51-54] It is usually (1) characterized by an area of steepening of at least 1 to 3 D in height and a diameter of at least 1 to 3 mm, (2) measured at least 1 month postoperatively, and (3) sometimes associated with monocular diplopia, haloes, and glare.[51] Many causes of central islands have been postulated; the exact reason is still unclear. The presence of an elevation in the corneal topographic map implies that there is an area of decreased ablation centrally. Surgeon, laser, patient-related, or a combination of these factors may be involved.[52] One plausible explanation for the cause of central islands is that the central cornea is relatively less ablated than the midperipheral cornea because it is better hydrated. Some of the laser pulses are rendered ineffective by the presence of more fluid in the corneal stroma centrally. Interestingly, central islands were detected only after the nitrogen gas effluent system was removed from certain excimer lasers. The nitrogen gas flow had the effect of removing moisture from the central cornea, thus improving excimer laser ablation of the central cornea. An "acoustic shock-wave" theory has been proposed by Maguen and Machat, who suggest that a flat laser beam, such as that of the VISX laser, drives stromal fluid centrally in the cornea, accounting for the relatively smaller ablative effect of the laser beam there.[51] It is useful to note that central islands have been reported mainly with the VISX laser and other lasers which utilize a flat beam profile, where the laser beam is homogeneous. It is essentially nonexistent with the Summit Excimed UV200 laser, which has a beam with a higher energy density centrally, compensating for the problem of excessive central corneal moisture.

The incidence of central islands varies, depending on the author, the ophthalmic center, and the time after PRK when corneal topography is performed. Lin analyzed the corneal topographic data on 502 consecutive eyes which had undergone PRK with the VISX excimer laser.[52] At the first postoperative month, four main morphological patterns of ablation were noted: "uniform" ablation in 12%, "keyhole" ablation in 44%, "semicircular" ablation in 18%, and "central islands" in 26%. However, the central islands tended to resolve with time, their incidence decreasing to 18% at 3 months, 8% at 6 months, and 2% at 12 months.[52] The authors support the concept of corneal remodeling after PRK, with progressive resolution of surface irregularities during the postoperative period. Similarly, Levin et al. believe that central island represent, in part, a biological phenomenon of healing rather than a purely mechanical effect of treatment.[53] In their study, they reported the incidence of central islands more than 3.00 D in height and 3.00 mm in diameter to be 13% at 3 months. In Maguen et al.'s study, the incidence of central island 3.00 mm in diameter and 3.00 D steep was 5% between 3 and 6 months postoperatively.[17] All eyes with such central islands underwent PRK without nitrogen flow. In 1.25% of eyes, these central islands persisted beyond 6 months. VISX has since modified its algorithm and introduced a new software program which adds additional pulses centrally to the laser beam to compensate for the problem of central islands.

ASTIGMATISM

Astigmatism, both regular and irregular, can occur in the cornea following PRK with the excimer laser. Irregular astigmatism may be induced in the immediate postoperative period as a result of epithelial healing and irregularity. Although some loss of BCVA may initially occur, this problem usually resolves over the next few weeks to months as the epithelium heals completely. Other, more serious causes of irregular astigmatism, such as decentration of the ablation zone, central islands, and the presence of nitrogen flow at the time of PRK, have already been described.

An increase in regular astigmatism after PRK for myopia is rather uncommon. There was no significant increase in regular astigmatism in many of the PRK studies mentioned above.[21,22,26] However, Goggin et al.[55] reported that 6 months following PRK, 36 of 60 eyes (60%) demonstrated a change in the cylindrical power or cylindrical axis of their refraction or both. The mean cylinder power change was 0.75 D, and in 15% of the eyes it was more than 1.00 D. The authors attributed the meridional change in refractive status to the corneal wound healing and remodeling process following excimer laser ablation.

CONCLUSION

PRK is a safe and effective procedure. Published reports indicate relatively predictable results in correcting low to moderate myopia with the excimer laser. Studies of prospective clinical trials of PRK vary in terms of the ranges of preoperative refractive error, follow-up times, and various treatment parameters used. As reported, these factors prevent direct comparison of clinical efficacy between studies. Tables 13.1 to 13.4 show marked variation among the studies for postoperative refraction within 1 D of emmetropia. In general, the precision of the refractive outcome is greatest with lower levels of preoperative myopia. Some of the more recent studies have achieved >90% refractive success (± 1.00 D) for corrections of up to −6.00 D. These results have led the FDA to approve PRK for up to −6.00 and −7.00 for the VISX and Summit laser systems, respectively. Further improvements in technique, modulation with pharmacological agents to control wound healing, and possible combination with automated lamellar keratoplasty may lead to more widespread use of PRK for the correction of myopia.

REFERENCES

1. Toboada J, Archibald CJ: An extreme sensitivity in the corneal epithelium to far UV ArF excimer laser pulses. In *Proceedings of Scientific Progress in Aerospace Medical Association*, San Antonio, 1981.
2. Trokel SL, Srinivasan R, Braren B: Excimer laser surgery of the cornea. *Am J Ophthalmol* 1985;100: 741–42.
3. McDonald M, Shofner S, Klyce S, et al: Clinical results of central photorefractive keratectomy (PRK) with the 193 nm excimer laser for the treatment of myopia: The blind eye study. *Invest Ophthalmol Vis Sci* 1989;30(Suppl):216.
4. SunderRaj N, Geiss MJ III, Fantes F, et al: Healing of excimer laser ablated corneas. An immunohistological evaluation. *Arch Ophthalmol* 1990;108:1604–1610.
5. Malley DS, Steinert RJ, Puliafito CA, Dobi ET: Immunofluorescence study of corneal wound healing after excimer laser anterior keratectomy in the monkey eye. *Arch Ophthalmol* 1990;108:1316–1322.
6. Hanna K, Pouliquen Y, Savoldelli M, et al: Corneal wound healing in monkeys 18 months after excimer laser photorefractive keratectomy. *Refract Corneal Surg* 1990;6:340–345.
7. Fantes FE, Hanna KD, Waring GO III, et al: Wound healing after excimer laser keratomileusis (photorefractive keratectomy) in monkeys. *Arch Ophthalmol* 1990;108:665–675.
8. McDonald MB, Frantz JM, Klyce SD, et al: One-year refractive results of central photorefractive keratectomy for myopia in the nonhuman primate cornea. *Arch Ophthalmol* 1990;108:40–47.
9. Taylor DM, L'Esperance FA Jr, Del Pero RA, et al: Human excimer laser lamellar keratectomy: A clinical study. *Ophthalmology* 1989;96:654–664.
10. McDonald MB, Frantz JM, Klyce SD, et al: Central photorefractive keratectomy for myopia. The blind eye study. *Arch Ophthalmol* 1990;108:799–808.
11. McDonald MB, Liu JC, Byrd TJ, et al: Central photorefractive keratectomy for myopia: Partially sighted and normally sighted eyes. *Ophthalmology* 1991;98: 1327–1337.
12. Seiler T, Kahle G, Kriegerowski M: Excimer laser

(193 nm) myopic keratomileusis in sighted and blind human eyes. *Refract Corneal Surg* 1990;6:165–173.
13. Gartry DS, Kerr Muir MG, Marshall J: Photorefractive keratectomy with an argon fluoride excimer laser: A clinical study. *Refract Corneal Surg* 1991;7:420–435.
14. Gartry DS, Kerr Muir MG, Marshall J: Excimer laser photorefractive keratectomy. 18 month follow up. *Ophthalmology* 1992;99:1209–1219.
15. Brancato R, Tavola A, Carones F, et al: Excimer laser photorefractive keratectomy for myopia: Results in 1165 eyes. *Refract Corneal Surg* 1993;9:95–104.
16. Kim JH, Hahn TW, Lee YC, Joo CK, Sah WJ: Photorefractive keratectomy in 202 eyes: One year results. *Refract Corneal Surg* 1993;9(Suppl 2):S11–S16.
17. Maguen E, Salz JJ, Nesburn AB, et al: Results of excimer laser photorefractive keratectomy for the correction of myopia. *Ophthalmology* 1994;101:1548–1556.
18. Hamberg-Nystrom H, Fagerholm P, Tengroth B, Epstein D: Photorefractive keratectomy for low myopia at 5 mm treatment diameter. A comparison of two excimer lasers. *Acta Ophthalmol* 1994;72:453–456.
19. O'Brart DPS, Gartry DS, Lohmann CP, Kerr Muir MG, Marshall J: Excimer laser photorefractive keratectomy for myopia: Comparison of 4.00 and 5.00 mm ablation zones. *J Refract Corneal Surg* 1994;10:87–94.
20. Heitzmann J, Binder PS, Kassar BS, Nordan LT: The correction of high myopia using the excimer laser. *Arch Ophthalmol* 1993;111:1627–1634.
21. Buratto L, Ferrari M: Photorefractive keratectomy for myopia from 6.00 to 10.00 D. *Refract Corneal Surg* 1993;9:(Suppl):S34–S36.
22. Sher NA, Hardten DR, Fundingsland B, et al: 193 nm excimer photorefractive keratectomy in high myopia. *Ophthalmology* 1994;101:1575–1582.
23. Talamo JH, Siebert K, Wagoner MD, et al: Multicenter study of photorefractive keratectomy for myopia of 6.00 to 8.00 D. *J Refract Surg* 1995;11:238–247.
24. Seiler T, Wollensak J: Myopic photorefractive keratectomy with the excimer laser. One year follow up. *Ophthalmology* 1991;98:1156–1163.
25. Tengroth B, Epstein D, Fagerholm P, et al: Excimer laser photorefractive keratectomy for myopia. Clinical results in sighted eyes. *Ophthalmology* 1993;100:739–745.
26. Ficker LA, Bates AK, Steele ADM, et al: Excimer laser photorefractive keratectomy for myopia: 12 month follow up. *Eye* 1993;7:617–624.
27. Epstein D, Fagerholm P, Hamberg-Nystrom Y, et al: Twenty-four month follow-up of excimer laser photorefractive keratectomy for myopia. Refractive and visual acuity results. *Ophthalmology* 1994;101:1558–1563.
28. Zadnick K, Mutti DO, Adams AJ: The repeatability of measurement of the ocular components. *Invest Ophthalmol Vis Sci* 1992;33:2325–2333.
29. Caubet E: Cause of subepithelial corneal haze over 18 months after photorefractive keratectomy for myopia. *Refract Corneal Surg* 1993;9(Suppl):S65–S70.
30. Gartry DS, Kerr Muir MG, Marshall J: The effect of topical corticosteroids on corneal haze following excimer laser treatment of myopia: An update. A prospective, randomised double-masked study. *Eye* 1993;7:584–590.
31. Fitzsimmons TD, Fagerholm P, Tengroth B: Steroid treatment of myopic regression. Acute refractive and topographical changes in photorefractive keratectomy patients. *Cornea* 1993;12:358.
32. Carones F, Brancato R, Venturi E, et al: Efficacy of corticosteroids in reversing regression after myopic photorefractive keratectomy. *Refract Corneal Surg* 1993;9(Suppl):S52–S56.
33. Fagerholm P, Hamberg-Nystrom H, Tengroth B, et al: Effect of postoperative steroids on the refractive outcome of photorefractive keratectomy for myopia with the Summit excimer laser. *J Cataract Refract Surg* 1994;29(Suppl):212–215.
34. Tavola A, Brancato R, Galli L, et al: Photorefractive keratectomy for myopia: Single vs double-zone treatment in 166 eyes. *Refract Corneal Surg* 1993;9(Suppl):S45–S52.
35. Scialdone A, Carones F, Bertuzzi A, et al: Randomized study of single vs double exposure in myopic PRK. *Refract Corneal Surg* 9(Suppl):S41–S43.
36. Seiler T, Derse M, Pham T: Repeated excimer laser treatment after photorefractive keratectomy. *Arch Ophthalmol* 1992;110:1230–123..
37. Epstein D, Tengroth B, Fagerholm P, et al: Excimer retreatment of regression after photorefractive keratectomy. *Am J Ophthalmol* 1994;117:456–461.
38. Kim JH, Sah WJ, Hahn TW, et al: Some problems after photorefractive keratectomy. *J Refract Corneal Surg* 1994;10(Suppl):S226–S230.
39. Seiler T, Schmidt-Petersen H, Wollensak J: Complications after myopic photorefractive keratectomy,

primarily with the Summit excimer laser. In Salz JJ (ed): *Corneal Laser Surgery.* St Louis, Mo, CV Mosby, 1995, pp 131–142.

40. Seiler T, Reckmann W, Maloney RK: Effective spherical aberration of the cornea as a quantitative descriptor in corneal topography. *J Cataract Refrac Surg* 1993;19(Suppl):155–165.

41. Lohmann CP, Fitzke FW, O'Brart DPS, et al: Haloes—a problem for all myopes? A comparison between spectacles, contact lenses and photorefractive keratectomy. *Refract Corneal Surg* 1993;9 (Suppl):S72–S75.

42. O'Brart DPS, Lohmann CP, Fitzke FW, et al: Night vision after excimer laser photorefractive keratectomy: Haze and haloes. *Eur J Ophthalmol* 1994;4: 43–51.

43. Uozato H, Guyton DL: Centering corneal surgical procedures. *Am J Ophthalmol* 1987;103:264–275.

44. Maloney RK: Corneal topography and optical zone location in photorefractive keratectomy. *Refract Corneal Surg* 1990;6:363–371.

45. Maguire LJ, Zabel RW, Parker P, et al: Topography and raytracing analysis of patients with excellent visual acuity 3 months after excimer laser photorefractive keratectomy for myopia. *Refract Corneal Surg* 1991;7:122–128.

46. Gimbel HV, Van Westenbrugge JA, Johnson WH, et al: Visual, refractive and patient satisfaction results following bilateral photorefractive keratectomy. *Refract Corneal Surg* 1993;9(Suppl 2):S5–S10.

47. Taylor HR, Guest CS, Kelly P, et al: Comparison of excimer laser treatment of astigmatism and myopia. *Arch Ophthalmol* 1993;111:1621–1626.

48. Cavanaugh TB, Durrie DS, Riedel SM, et al: Topographical analysis of the centration of excimer laser photorefractive keratectomy. *J Cataract Refract Surg* 1993;19(Suppl):136–143.

49. Klyce SD, Smolek MK: Corneal topography of excimer laser photorefractive keratectomy. *J Cataract Refract Surg* 1993;19(Suppl):122–130.

50. Lin DT, Sutton HF, Berman M: Corneal topography following excimer photorefractive keratectomy for myopia. *J Cataract Refract Surg* 1993;19(Suppl): 149–154.

51. Maguen E, Machat JJ: Complications of photorefractive keratectomy, primarily with the VISX excimer laser. In Salz JJ (ed): *Corneal Laser Surgery.* St Louis, Mo, CV Mosby, 1995, pp 143–158.

52. Lin DTC: Corneal topographic analysis after excimer photorefractive keratectomy. *Ophthalmol* 1994;101: 1432–1439.

53. Levin S, Carson CA, Garrett SK, et al: Prevalence of central islands after excimer laser refractive surgery. *J Cataract Refract Surg* 1995;21:21–26.

54. Parker PJ, Klyce SD, Ryan BL, et al: Central topographic islands following photorefractive keratectomy. *ARVO Abstracts. Invest Ophthalmol Vis Sci* 1993;34:803.

55. Goggin M, Algawi K, O'Keefe M: Astigmatism following photorefractive keratectomy for myopia. *J Refract Corneal Surg* 1994;10;540–544.

CHAPTER 12

PTK in the Management of PRK Complications

Dimitri T. Azar, Walter J. Stark, Roger F. Steinert

Phototherapeutic keratectomy (PTK) is valuable in the surgical management of photorefractive keratectomy (PRK) complications including central islands, undercorrection, anterior stromal haze, and decentration. These complications can be minimized using careful surgical technique and may respond to medical therapy, but occasionally require surgical intervention. In this chapter we will discuss the surgical management of such PRK complications. In all cases, the risks of early surgical intervention should be weighed against the likelihood of spontaneous resolution.

CORNEAL TOPOGRAPHY AFTER PRK

Several characteristic topographical patterns of ablation following PRK have been observed using subtraction maps. In a study of corneal topography following PRK, Lin[1] reported uniform semicircular, keyhole, and central island patterns. A smooth or uniform ablation pattern was most commonly observed, showing concentric flattening of the ablation zone with the central area being flatter than the periphery (Figure 12.1A). A semicircular ablation pattern was observed in 18% of the eyes, showing a power difference of 1.50 D or more across meridians relative to the center of the ablation zone (Figure 12.1B). A keyhole nonuniform ablation pattern with a power difference of 1.50 D or more between equidistant points on any meridian relative to the center of the ablation zone was seen in 12% of eyes. A central island pattern of ablation, present in the remaining 26% of eyes, was defined as a 1.50 D or more increase in the central power of the ablation zone and occupying at least 2.50 mm of the central cornea (Figure 12.1C and 12.1D). Eyes with central island topographic pattern of ablation have also been observed to demonstrate a greater loss of best corrected visual acuity compared with other ablation patterns at the first postoperative month.

In another study of corneal surface irregularities following PRK, Hersh et al.[2] defined seven topographical ablation patterns as follows: (1) homogeneous, or uniform flattening; (2) smooth toric bowtie with axis, which represents a symmetric ablation zone with a greater flattening in the steep preoperative axis; (3) smooth toric bowtie against axis in which the greater flattening is in the flat preoperative axis; (4) generalized irregularities over the ablation zone with more than one area measuring more than 0.5 mm and a difference of more than 0.50 D in power from other areas; (5) keyhole or semicircular, when a topographic region measuring at least 1.0 mm and 1.00 D of relatively less flattening extends in from the periphery (keyhole) or a foreshortening of the ablation zone in one meridian (semicircular); (6) central island, a central area of relatively less flattening occupying at least 1.0 mm and measuring 1.00 D or more in power; and (7) a generally homogeneous pattern with irregularities measuring less than 1.0 mm or 1.00 D in power.

Laser companies have modified their laser algorithms to minimize nonuniform ablation patterns after PRK. Although this may have resulted in improvements of best corrected visual acuity, topography is still valuable in identifying patients with central islands and decentered treatments requiring surgical intervention.

MANAGEMENT OF CENTRAL ISLANDS AFTER PRK

Various studies have shown that most central islands occur immediately following PRK. As with subepithelial haze, there is a steady decrease in cen-

Figure 12.1. *A:* Uniform PRK ablation showing preoperative topography (top left), postoperative topography (bottom left) and difference map (right). *B:* Semicircular ablation, *C:* Central island, *D:* Astigmatic central island treatment compounding preoperative with the rule astigmatism (top left), resulting in increased postoperative astigmatism (bottom left).

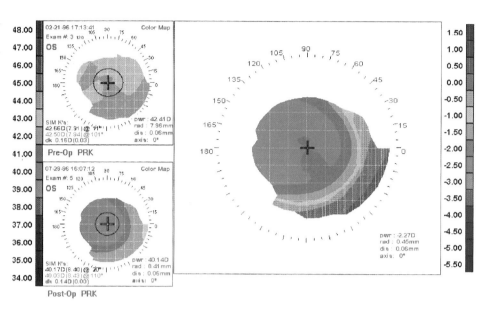

A

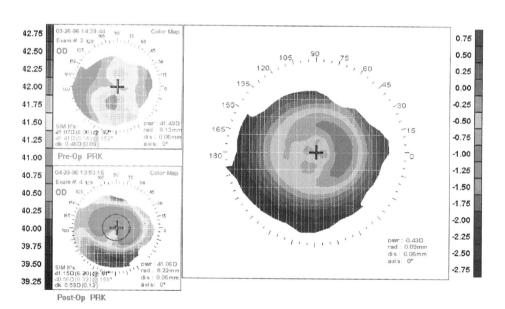

B

PTK in the Management of PRK Complications

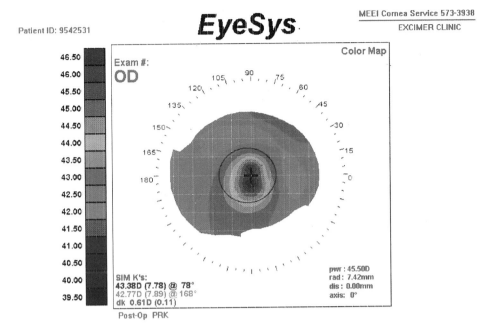

C

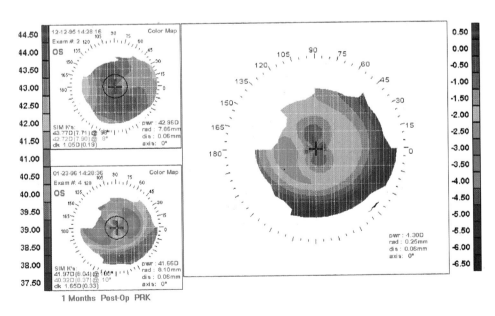

D

tral island severity between 3 and 12 months postoperatively.[3-7] Central islands represent local steepening of the central cornea as compared to surrounding paracentral areas. They have been reported following keratomileusis, automated lamellar keratoplasty, and radial keratotomy,[8] but are most commonly observed in patients with moderate to high myopia who undergo large area 6-mm PRK treatments.[1,9] Newer PRK treatment algorithms in certain lasers have incorporated central island protocols to minimize central islands.

The incidence of central islands varies. Lin has observed central islands in 26% of eyes at 1 month after PRK and noted that most central islands tended to resolve with time.[1]

Pathogenesis of Central Islands

Central islands may occur due to one or more of the following factors[10]: (1) nonhomogeneous beam intensity of the excimer laser, especially after prolonged use whereby the center of the laser mirrors receives constant irradiation with subsequent relative reduced transmission of the laser energy before it strikes the cornea; (2) interference with the central portion of the laser beam by the plume of ejected tissue fragments; (3) variability of corneal hydration during the procedure with the central area retaining greater hydration;[11] (4) epithelial removal resulting in residual epithelial cells in the central area of the cornea; (5) greater amounts of tissue removal and delayed wound closure of the central area which may lead to greater tissue response centrally and increased relative epithelial remodeling in the deeper areas of the treatment zone, in turn resulting in relative steepening of the central cornea; (6) postoperative exposure to drugs of the central cornea prior to epithelial closure; (7) treatment decentration and intraoperative treatment drift resulting in more tissue ablations in the paracentral areas with relative central island formation; (8) ectasia of the central cornea after deep excimer ablations; (9) greater degree of epithelial hyperplasia in the central cornea; and (10) an aberration of the software used in videokeratoscopic analysis of the topographical images.

The variability of central islands and of subepithelial haze are related to the variability in wound healing responses of individual patients. In a study of wound healing response after PRK in rhesus monkey corneas, Fantes et al.[3] demonstrated transient anterior stromal fibroplasia with production of new extracellular matrix, mainly type III collagen, type VII collagen,[12] and keratan sulfate[3,12] which was associated with the clinical observation of haze. There was a strong correlation between the clarity of healed cornea and the absence of new collagen and epithelial hyperplasia.[13] (Figure 12.2)

Clinically, these various patterns of wound healing are seen in different regions of the cornea in the same patient, showing areas of excessive local subepithelial collagen deposition in one region and minimal, if any, in another. When the excessive haze occurs predominantly in the central cornea, central islands may be observed. (Figure 12.2)

Risk Factors

The most important risk factor for the development of central islands is the degree of intended myopic correction. Higher myopic and higher cylindrical corrections require greater treatment depths, and any imperfection in the treatment contour may increase the chance of central islands.

With proper surgical techniques, smoother wound contours, and optimal intraoperative stromal hydration, the risk of corneal island formation may be reduced.[5,14] The use of a larger optical zone increases the risk of central island formation, if central island protocols are not incorporated into treat-

Figure 12.2. Subepithelial haze associated with central island.

ment algorithms. Slow reepithelialization and persistent epithelial defect[5] may predispose to central island and postoperative haze formation.[15] Central islands were non-existent with 5.0 mm optical zone PRK.

In the VISX (VISX, Incorporated, Santa Clara, CA) studies, nitrogen flow discontinuation decreased the likelihood of subepithelial haze formation,[5] but has been shown to increase the chance of central island formation. It is expected that by increasing the turnout of the plume of ejected tissue fragments, the central cornea receives more homogeneous laser energy, thus reducing the chance of central island formation.

Management

Central islands[16] and corneal haze[17] have been observed to resolve spontaneously in most patients by 6 months postoperatively. Although steroids can be used to control haze, they cannot eliminate all haze or scars. In addition, there is no evidence whether topical steroids have any effect on central islands. O'Brart et al.[14] observed that no patients following a 6.00-mm-zone PRK ablation had central island patterns at 12 months postoperatively even though one patient had a small central island at the 1-, 3-, and 6-month follow-up examinations. The central island was not present at 12 months and 20/20 vision was achieved.

When visually significant central islands persist without improvement beyond 6 to 12 months, despite medical interventions, surgical retreatment may become necessary. In a study of 645 eyes by Snibson et al., seven (12%) of the retreated patients had central islands identified on videokeratography following the original PRK treatment. Results showed no additional haze could be attributed to the deeper stromal ablation upon retreatment. It is advised that retreatment be avoided for at least 6 months postoperatively[18] to demonstrate refractive stability over 3 months after discontinuation of topical corticosteroids before retreatment and allow time for spontaneous improvement.[18]

The objective of retreatment is to minimize central islands, eliminate the associated refractive error, and restore topography. A transepithelial approach, using combined PRK/PTK modality, in which the epithelium is ablated with a PTK approach, is preferred. A key to retreatment is to calculate the power difference of the central island from corneal 1 month post-laser topographical maps. Ideally, the optic zone should be at least as large as the central island. To begin, at least 200 pulses or 50 μm of a large area (6.0–6.5 mm diameter) epithelial ablation is programmed. The illumination of the microscope is set at the lowest possible level allowing adequate visualization of the area of treatment. Epithelial treatment is halted once a dark area of stromal ablation is noted (loss of epithelial fluorescence). When excessive haze is still evident by the slit lamp examination, an additional 5 to 15 μm PTK can be ablated.[19] Alternatively, following the ablation of the epithelium, the PRK treatment is based on the calculated depth of ablation using height data obtained from corneal topography. The principle is to carefully monitor the depth of ablation not to exceed the estimated height of the island. In the United States, the FDA only allows PRK algorithms to be operated at a 6.0 mm optical zone, which is unsatisfactory for correction of a central island. The solution is to use the PTK mode set at the maximum diameter of the central island. The number of pulses is calculated by the Munnerlyn Formula

$$t = \frac{S^2 D}{3}$$ where t = central thickness of tissue removed
S = ablation diameter
D = dioptric correction

For example, an island with a 3 mm diameter and a 40 elevation requires 12 micron ablation. At approximately 0.25 microns of tissue removed per pulse, 48 pulses are needed.

The importance of prevention of central islands cannot be overemphasized. PRK in higher myopes should be preceded or accompanied by "central island treatment protocol." This is automatically incorporated in newer large area lasers. Alternative approaches, such as laser assisted in situ keratomileusis (LASIK), should be considered. It is not clear if central island pretreatment protocols are necessary in LASIK. As in PRK, spontaneous improvement in central islands after LASIK has been observed. In performing PRK retreatment for central islands, it is important to control the overall intraoperative hydration status of the stroma. Likewise, control of dry eyes in the immediate postoperative period may be helpful in some patients.[20] Early steroids intervention and epithelial healing can also help minimize the risks of central island treatment. Retreatment of the apex of the island is crucial to avoid irregular topographical outcomes and exaggeration of the central island. Finally, it is just as important, if not more, to keep in mind that unrealistic preoperative expectations of central island retreatment should be avoided.

MANAGEMENT OF PRK UNDERCORRECTIONS

Retreatments can be performed for undercorrections with or without coexisting corneal haze. Undercorrected PRK not associated with visually significant subepithelial haze may be due to one or more of the following factors:

1. Abnormally low ablation rate
2. Intentional undercorrection, especially in high myopia
3. Planned monovision
4. Regression of initial refractive result

Retreatment is usually successful and has minimal side effects. After extensive discussions with the patient, many surgeons intentionally undercorrect in order to avoid the problem of overcorrection, especially in presbyopic or prepresbyopic patients. The patient is informed about the risks of retreatment which are weighed against the risks of overcorrection of initial treatment. Younger patients may wish to proceed with full PRK treatment while older patients, concerned about loss of near vision, may prefer intentional undercorrections. The management of patients with PRK undercorrections not associated with visually significant haze is less complicated than for those patients with associated subepithelial haze.

There is ordinarily no significant loss of spectacle corrected visual acuity, and corneal topography shows homogenous ablation zone without decentration. Management should be modified if topographical or slit lamp abnormalities are noted.

The timing of PRK retreatment is usually based on the extent of undercorrection. Patients with slight undercorrection may be initially dissatisfied with the outcome of surgery, but may subsequently be satisfied with monovision if the second eye achieves full correction. Patients' expectations have to be lowered when retreatment is considered for undercorrections of <0.75 D. These patients may also suffer from undiagnosed surface irregularities, glare, halos, double vision, and residual astigmatism that may persist after retreatment.

Two surgical approaches of retreatment have been successfully employed. In the first, the epithelium is scraped and the residual error is corrected using PRK retreatment on the stromal ablation bed. This first method assumes that epithelial thickness following PRK will remain unchanged after retreatment. The second method assumes an average power of 0.75 D for the epithelial layer after PRK. A transepithelial treatment is performed using 6.0–6.5 mm diameter PTK as described above for central island retreatment. The endpoint of epithelial retreatment is loss of epithelial fluorescence in the periphery of the ablation zone. An additional treatment of up to −0.75 D is added to the amount of intended correction and programmed into the laser for PRK retreatment. In the absence of the peripheral dark ring of fluorescence, one must avoid programming the additional 0.75 D treatment.

MANAGEMENT OF CORNEAL SUBEPITHELIAL HAZE AFTER PRK

Corneal subepithelial haze is commonly seen following PRK. It is described as loss of corneal transparency which may arise from scattering of light incident to the ablated zone. Corneal haze is part of the normal corneal healing response after PRK.[3] Corneal haze peaks at 1–6 months following PRK. There is a steady decrease in severity up to 18 months postoperatively.[4,5,7] Transient haze, such as this, typically has a "reticular" appearance. Scarring, however, is defined as a localized or diffuse persistent opacity which is often associated with visual symptoms and topographic irregularities.

The degree of haze varies between patients. In their study of 24-month follow-up of excimer laser for myopia, Epstein et al.[21] observed that 96% of the eyes had a subepithelial haze of 1+ or less, while 3% of the eyes had a haze of 2+, and 1% with haze of 3+. In another study by Thompson et al.,[22] 89.3% of all patients were free of clinically significant haze 12 months after surgery, and no eyes suffered from marked haze. There was a tendency for early postoperative haze to diminish over time. Moderate to severe haze was found in 3.1% of cases at 3 months postoperatively and dropped to 2.6% at 12 months postoperatively.[22] Seiler et al.[23] observed that the incidence of manifest scars after PRK depends on the attempted correction. In correction of the lower myopia group (≤ -3.0 D), 0% had scar, whereas 1.1% and 17.5% of cases had manifest scars in correction of the moderate myopia group (-3.1 to -6.0 D) and the high myopia group (-6.1 to -9.0 D), respectively.

As discussed, the variability of haze is related to the variability in wound healing responses of individual patients. In a study of wound healing response after PRK in rhesus monkey cornea, Fantes et al.[3] demonstrated that the epithelium healed well without undergoing significant hyperplasia. However, there was a transient anterior stromal fibroplasia with production of new extracellular matrix, mainly type III collagen, type VII collagen,[12] and keratan sulfate[3,12] which was associated with the clinical observation of haze. There was a strong correlation between the clarity of healed cornea and the absence of new collagen and epithelial hyperplasia.[24]

At the epithelial–stromal junction, areas of basement membrane discontinuity persist for 6 months.[25–27] Discontinuity of these basement membranes was typically seen with epithelial healing over bare stroma. The irregularities in the basal epithelial cells, the vacuolation between lamellae, increased number of fibroblasts, and production of extracellular matrix as a consequence of epithelial–stromal wound healing cause light scattering and corneal haze.[3,27–30]

The changes in anterior stroma (subepithelial stromal haze) appear at 2–3 weeks after ablation and peak at 6–8 weeks. During this time, activated fibroblasts migrate into the existing normal structures and secrete new extracellular matrix. After approximately 3 months, the number of fibroblasts decrease, and extracellular matrix is remodeled, which results in regression of haze.

Investigators have found various structural changes in response to PRK and corneal wounds. Malley demonstrated the presence of type VII and type III collagen in monkey corneas by indirect immunofluorescence microscopy 18 months after PRK; there were also fibronectin and laminin deposited in high quantities, which was associated with an increase in size and number of keratocytes at the treatment area.[31] Hanna et al. described early histochemical changes in rabbits, such as increased amount of type IV collagen, proteoglycans, fibronectin, and laminin.[32] Azar et al.[27] found that after excimer laser, the matrix metalloproteinase MMP-9 was expressed in corneal epithelium and stroma during wound closure. Another collagenase, MMP-2, was present in stroma before wounding and increased following wounding. It is possible that these proteolytic enzymes play a role in stromal remodeling in normal cornea and in scar formation and clearing postexcimer wound.[27]

Durrie reported three principal patterns of healing response following PRK in human patients: normal healers were those with a small amount of hyperopia that faded slowly over time; inadequate healers had minimal haze and remained hyperopic; and aggressive healers developed significant regression with subepithelial haze.[33]

Risk Factors of Corneal Subepithelial Haze

The most important risk factor for the development of subepithelial scarring is the degree of myopic correction. Higher myopic and higher cylindrical corrections require greater treatment depths, which may explain the increased degree of subepithelial haze. Caubet observed that at 1 month, mean clinical anterior stromal haze was 0.5, 0.74, and 1.15 in the -2, -4, and -6 D myopic correction, respectively.[6] At 3 months, mean haze was 0.73, 1.13, and 1.62; and at 18 months, they were 0.07, 0.17, and 0.83 in the -2, -4, and -6 D myopic corrections, respectively. Oral contraceptives have been noted to increase the risk of scarring after PRK by a factor of 14.

With proper surgical techniques, smoother wound contours,[14] and optimal intraoperative stromal hydration,[5] the risk of corneal haze formation can be minimized. The use of a larger optical zone (>4.5 mm) give markedly less haze after PRK.[14] O'Brart et al. investigated the effect of ablation diameter, depth, and edge contour on the outcome of PRK and showed that haze was significantly reduced at 1, 3, and 6 months in the 6.00-mm treated eyes. A multizone ablation did not yield different from that of single zone. While uncontrolled clinical reports claim benefit from multizone ablations, this has not been demonstrated in a controlled study.

The extent of haze has also been found to be associated with younger patient age,[7] the male sex,[6] nitrogen flow,[5] rapid steroids withdrawal,[6] keloid former,[34] pregnancy,[34] and dry eyes.[35] Medications like acutane, amiodarone, estrogen,[38] and progesterone[38] may contribute to haze formation. Oral contraceptive use may have to be avoided in order to minimize scar formation. Untreated allergic conjunctivitis has also been identified as another potential risk factor for haze and regression following PRK.[36]

Furthermore, history of previous refractive surgeries, such as radial keratotomy, astigmatic keratectomy, and cataract surgery, may also predispose to postoperative haze development. In a multicenter trial of PRK for residual myopia after previous ocular surgery, Maloney et al.[37] observed mild haze in 11 of 107 eyes, moderate haze in 5 of 107 eyes, and marked haze in 3 of 107 eyes, which represented a greater incidence of haze than in PRK without previous surgery. They concluded that the previous corneal surgery may have increased scar formation due to fibroblast activation.

Medical Management

Subepithelial haze following PRK is usually associated with regression of the initial refractive effect. Persistent haze after 6 months postoperatively can be managed either medically or surgically. Medical and surgical management decisions are usually based on the presence and extent of associated regression of the refractive effect.

Topical steroids are often prescribed in the first 3 months following PRK, especially for myopic corrections of >3 D, and if nonsteroidal anti-inflammatory drugs (NSAID) are used. Although steroids can be used to control haze, they cannot eliminate haze.[35] In a study of 10 blind and 13 sighted eyes, Seiler et al.[38] showed that postoperative steroids can minimize corneal haze. In a rabbit model, postoperative steroid treatment was shown to reduce both stromal haze and thickness of subepithelial new collagen layer.[39] However, another study showed that steroid treatment was associated with persistence of stromal thinning.[40] In addition, it was observed that the extent of anterior stromal haze was not influenced by the presence or absence of steroids at any stage.[21] Therefore, although steroids have been shown to be helpful in controlling haze development, their long-term benefits remain uncertain. The risks of steroid complications, such as steroid-induced glaucoma, posterior subcapsular cataract, ptosis, increased risk of infection, and rebound upon withdrawal, should certainly be taken into consideration.

Topical NSAID such as diclofenac and ketorolac have been shown to decrease corneal haze in rabbits in the early postoperative period.[34,42] However, as NSAIDs delay epithelial wound healing, they may actually increase haze formation, which has led many investigators to limit their role to pain relief, and avoid using them for more than 24–28 h.[43]

Other potential medications that may have a role in the management of corneal scarring after PRK include: mitomycin C,[23] antioxidants,[44] MMP inhibitors,[26] plasmin- and plasminogen activator inhibitors,[45] and alpha 2b-interferon.[46] These drugs, however, are still experimental.

Surgical Management

When visually significant haze persists for more than 6 months despite medical interventions, surgical retreatment may become necessary. The goal of the retreatment is to achieve trace haze, and not necessarily a clear cornea. Achieving a clear cornea may require overtreatment and subsequently result in significant hyperopia. The objective is to debulk haze, or transform it from clinically significant to insignificant haze, eliminate the associated refractive error and restore regular topography. As in treatment of central islands, a transepithelial approach, using a combined PRK/PTK modality, in which the epithelium is ablated with a PTK approach, is typically the best approach in cases of clinically significant haze. A key to retreatment is to have a maximal optical zone. Ideally, the optic zone should be at least as large as the primary procedure. A starting point of two hundred pulses or 50 μm of epithelium ablation is usually recommended.[19] The goal or the endpoint of retreatment should be about 90% of haze clearance, leaving the cornea with a mottled "stromal" surface. As stated it is ideal to leave a trace amount of residual haze avoiding overtreatment and induced hyperopia. When excessive haze is still evident by the slit lamp examination, an additional 5 to 15 μm can be ablated.

Alternatively, following the ablation of the epithelium, a PRK approach can be used. Similarly, the principle is to carefully monitor the depth of ablation. Ideally, only 50–60% of associated myopia is treated to avoid full correction.[19] In conclusion, in retreatment the emphasis should be placed on the depth of ablation rather than the diopters corrected in order to avoid overtreatment.

Other surgical approaches in the management of persistent haze, such as corneal scraping, homoplastic automated lamellar keratoplasty, and penetrating keratoplasty, should rarely be considered.

In order to prevent haze from developing, PRK in high myopes should be avoided if at all possible. Alternative approaches, such as LASIK, should be considered[47] in cases of very high myopia and for the fellow eye of patients developing moderate haze or scarring following PRK.

MANAGEMENT OF PRK DECENTRATIONS

Treatment decentration has been associated with reduced visual function following PRK. When the ablation zone is decentered from the entrance pupil, unwanted complications such as glare, edge effects, and refractive errors may result.[49,62] Computerized videokeratography has been utilized in the postoperative evaluation of PRK centration, allowing an objective measurement of surgical results and location of the treatment zone.[1-5, 50-53] (Figure 12.3A).

Two methods of analyzing corneal topographical information have been described, namely axial and tangential (instantaneous) topography. The latter provides instantaneous radius of curvature data which are valuable in detecting the edge of PRK treatments with greater precision than axial topography.[54-56] The theoretical relationship between the axial and tangential (instantaneous) power is that the axial power at a specific point in the cornea is equal to the average of the instantaneous powers from the axis to that point of interest.[56]

In postoperative evaluation of PRK decentration, axial topographical maps have been commonly used to compare the distance between the pupillary center and the area of maximal ablation. We have used a method of determining the edge of the ablation zone by using tangential topographical maps, and deriving the center of the treatment zone, which allowed us to compare treatment decentration and displacement (shift) to treatment drift.

Amano et al.[53] observed a tendency of downward decentration and suggested that it may have been a result of the reflex upward movement of the eyeball as in Bell's phenomenon secondary to the topical anesthesia during the treatment. In addition, it has been demonstrated that pupils tend to shift with miosis, most commonly superior nasally.[57]

Eccentric ablations have been shown to result in postoperative halos, increased glare, and refractive errors, but there is a lack of consistency in the literature regarding the correlation between the best corrected acuity and decentration. Some authors[1] found no correlation between the centration data and the best corrected visual acuity of the patients, whereas others[49,52,58-62] noted a decrease in visual acuity with decentration. We found no statistically significant correlation between laser shift and the best corrected visual acuity.[63] Instead, we observed that the drift index was better correlated with the postoperative visual acuity outcomes. Drift occurs when the eye moves involuntarily during treatment, or when the surgeon attempts to correct apparent decentration during treatment. The result is an uneven ablation area with flatter treatment zone shifted peripherally, leaving the central area of the ablation zone with a higher corneal surface power difference.[63] Consequently, the maximum power of ablation moves away from the center of the ablation zone, resulting in uneven, and "undercorrected" corneal surface over the pupillary axis. This non-uniformity and steeping secondary to drift in the central cornea is responsible for the loss of best corrected visual acuity in patients who had near perfect laser centration of the treatment (no treatment displacement).

In the management of treatment decentrations, it may be useful to make the distinction between the effects of laser drift and shift. While laser shift is a direct consequence of misalignment of laser beam with the center of the entrance pupil from the beginning of treatment, laser drift occurs with any subtle movement of the eyeball irrespective of centration, resulting in a relatively nonuniform distribution of surface powers within the treated zone.[63] Figure 12.4A and 12.4B demonstrate the effect of drift on the best corrected visual acuity. Both eyes were equally mildly decentered relative to the center of the pupil (0.31 mm); they were also both

Figure 12.3. *A:* Preoperative topography showing decentered treatment. The patient underwent PTK treatment of the epithelium centered around the pupillary area followed by PRK treatment of the residual refractive error. *B:* Postoperative topography showing recentration of ablation zone.

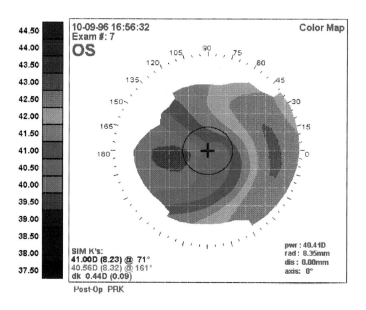

A

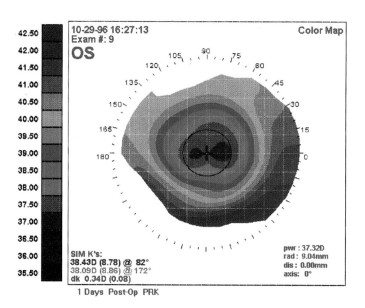

B

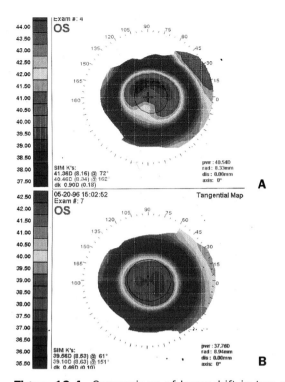

Figure 12.4. Comparison of laser drift in two patients having similar amounts of treatment displacement (shift). *A*: Tangential topography showing a laser drift effect in the superior direction. The treatment, with an intended myopic correction of −6.20 D, was slightly shifted inferotemporally (r = 0.31 mm). Note the area of greatest ablation (blue) was drifted upward, resulting in a nonuniform central ablation power. The change in central ablation power in the central 4 mm² relative to the pupillary center was 3.00 D, while the arc of the second flattest area was 3.07 rad. The shortest distance from the center of ablation to the flattest area was 1.00 mm. The drift index was 0.98. The best corrected vision 1 month following PRK was 20/40. *B*: Tangential topography of a left eye with an intended myopic correction of −5.50 D with similar degree of shift as the map in Figure 12.4A (0.31 mm). Compared to Figure 12.4A, the central power is more homogeneous without gross drift effect (drift index = 0.03). 20/20 visual acuity was achieved. (From Azar DT, Yeh P; reference 63).

shifted inferotemporally. However, the eye in Figure 12.4A experienced a greater drift effect with a drift index of 0.98, resulting in 20/40 visual acuity. As a result of downward intraoperative drift, the flattest ablation zone moves upward in the periphery, leaving the central optic zone relatively undercorrected with an uneven distribution of surface powers across the central cornea. Figure 12.4B, on the contrary, shows a fairly uniform central ablation area with the flattest area within treated central pupillary zone. In this case, the drift index is found to be minimum (0.03), associated with 20/20 vision.[63]

Four centration/decentration patterns may occur after PRK: (1) good initial centration with no eye movement during treatment; (2) good initial centration with involuntary eye movement; (3) poor initial centration with no eye movement; (4) poor initial centration with correctional eye movement.[63] One would expect that the following would occur: in the first situation, low shift and low drift would be expected; in the second situation, high shift and high drift; in the third situation, high shift and low drift; and in the fourth situation, low shift and high drift. Low-shift and low-drift and high-shift, low-drift patients gave the best mean best corrected visual acuity, whereas the groups with high drift had the worst mean best corrected visual acuity. Therefore, PRK retreatment may be helpful in patients with laser drift especially when associated with reduced visual outcomes. In addition, given that good visual outcome was achieved in decentered treatments without drift, it may be advisable to consider retreatment only if a treatment edge effect is noted resulting in glare and halos.[63]

The surgical retreatment technique that we recommend for patients with low drift (high shift and edge glare) is transepithelial treatment with recentration of PTK and PRK treatment on the center of the entrance pupil.[61,62] The initial PTK treatment is continued through the epithelium. Low illumination allows the surgeon to observe the pattern of epithelial fluorescence start disappearing in the peripheral (previously untreated) zone and proceed until the center of the cornea is reached. Subsequent PRK treatment is performed to correct the residual refractive error (Fig. 12.3A and B).

In patients with high drift during treatment, the PTK treatment (centered over the center of the entrance pupil) is followed by application of masking fluids, as described in Chapters 7 and 8. This is carried on until the surface of the stroma is smooth prior to PRK treatment of the residual refractive error.

Without a doubt, consistent and proper centration of the excimer laser is critical in achieving optimal PRK results. Moreover, computerized corneal topography continues to provide essential and objective information about centration and corneal contour following PRK. Our method using tangential topographical maps seems to be valuable in determining the edges and center of PRK ablation,[63] and seems to indicate that laser drift may be a more important determinant of postoperative visual acuity following PRK.

REFERENCES

1. Lin DTC: Corneal topographic analysis after excimer photorefractive keratectomy. *Ophthalmology* 1994; 101:1432-1439.
2. Hersh PS, Schwartz-Goldstein BH: The Summit photorefractive keratectomy topography study group. Corneal topography of phase III excimer laser photorefractive keratectomy: characterization and clinical effects. *Ophthalmology* 1995;102:963-978.
3. Fantes FE, Hanna KD, et al: Wound healing after excimer laser keratomileusis (photorefractive keratectomy) in monkeys. *Arch Ophthalmol* 1990;108:665-675.
4. Stein HA, Cheskes A, Stein RM: Complications and their management. In *The Excimer, Fundamentals and Clinical Use*. Thorofare, NJ, SLACK Inc, 1996.
5. Maguen E, Machat JJ: Complications of photorefractive keratectomy, primarily with the VISX Excimer Laser. In Salze JJ (ed): *Corneal Laser Surgery*. St Louis MO, Mosby, 1995; pp 143-158.
6. Caubet E: Cause of subepithelial corneal haze over 18 months after photorefractive keratectomy for myopia. *Refract Corneal Surg* 1993;9(Suppl):S65-70.
7. Gartry DS, Muir MGK, Marshall J: Excimer laser photorefractive keratectomy, 18-month follow-up. *Ophthalmology* 1992;99(8):1209-1219.
8. Price FW: Central islands of corneal steepening after automated lamellar keratoplasty for myopia. *J Refract Surg* 1996;12:36-41.
9. Wilson SE, Klyce SD, McDonald MB, et al: Changes in corneal topography after excimer laser photorefractive keratectomy for myopia. *Ophthalmology* 1991;98:1338-1347.
10. Grimm B, Waring III GO, Ibrahim O: Regional variation in corneal topography and wound healing following photorefractive keratectomy. *J Refract Surg* 1995;11:348-357.
11. Ousley PJ, Terry MA: Hydration effects on corneal topography. *Arch Ophthalmol* 1996;114:181-185.
12. SundarRaj N, Geiss MJ, Fantes F: Healing of excimer laser ablated monkey corneas. An immunohistochemical evaluation. *Arch Ophthalmol* 1990; 108:1604-1610.
13. Goodman GL, Trokel SL, Stark WJ: Corneal healing following laser refractive keratectomy. *Arch Ophthalmol* 1989;107:1799-1803.
14. O'Brart DPS, Corbett MC, Verma S: Effects of ablation diameter, depth and edge contour on the outcome of photorefractive keratectomy. *J Refract Surg* 1996;12:50-60.
15. Machat JJ: Postoperative PRK patient management. In *Excimer Laser Refractive Surgery, Practice and Principles*. Thorofare, NJ, SLACK Inc, 1996.
16. Krueger RR, Saedy NF, McDonnell PJ: Clinical analysis of topographic steep central islands following excimer photorefractive keratectomy (PRK). ARVO Abstracts. *Invest Ophthalmol Vis Sci* 1994; 35(4, Suppl):1740.
17. Carson CA, Snibson GR, Taylor HR: An analysis of clinical correlations one, three, six, and 12 months after excimer laser photorefractive keratectomy. *Lasers Light Ophthalmol* 1994;6:249-257.
18. Snibson GR, McCarty CA, Aldred GF, et al: Retreatment after excimer laser photorefractive keratectomy. *Am J Ophthalmol* 1996;121:250-257.
19. Machat JJ: PRK retreatment techniques and results. In *Excimer Laser Refractive Surgery, Practice and Principles*. Thorofare, NJ, SLACK Inc, 1996.
20. Tervo T, Mustronen R, Tarkkanen A: Management of dry eye may reduce haze after excimer laser photorefractive keratectomy [Letter]. *Refract Corneal Surg* 1993;9(Suppl):306.
21. Epstein D, Fagerholm P, Namberg-Nystrom H, et al: Twenty-four month follow-up of excimer laser photorefractive keratectomy for myopia. *Ophthalmology* 1994;101(9):1558-1564.
22. Thompson KP, Steinert RF, Daniel J, et al: Photorefractive keratectomy with the Summit Excimer Laser: The Phase III U.S. Results. In: *Corneal Laser Surgery*. St Louis, MO, Mosby, 1995.
23. Seiler T, Holschback A, Derse M, et al: Complications of myopic photorefractive keratectomy with the excimer laser. *Ophthalmology* 1994;101(1):153-160.
24. Goodman GL, Trokel SL, Stark WJ: Corneal healing following laser refractive keratectomy. *Arch Ophthalmol* 1989; 107:1799-1803.
25. Fountain TR, de la Cruz Z, Green WR, Stark WJ, Azar DT: Reassembly of corneal epithelial adhesion

structures after excimer laser keratectomy in humans. *Arch Ophthalmol* 1994;112:967–972.
26. Azar DT: Epithelial and stromal wound healing following excimer laser keratectomy. *Semin Ophthalmol* 1994;9(2): 102–105.
27. Azar DT, Chamon W, Stark WJ, et al: Matrix metalloproteinase expression in laser-ablated corneas. *Invest Ophthalmol Vis Sci* 1993;34(Suppl):704.
28. Ramirez-Florez S, Maurice DM: Inflammatory cells, refractive regression, and haze after excimer laser PRK. *J Refract Surg* 1996;12:370–381.
29. Talamo JH, Gallamudi S, Green R, et al: Modulation of corneal wound healing after excimer laser keratomileusis using topical mitomycin C and steroids. *Arch Ophthalmol* 1991;109:1141–1146.
30. Marshall J, Trokel SL, Rothery S: Long-term healing of the central cornea after photorefractive keratectomy using an excimer laser. *Ophthalmology* 1988; 95(10):1411–1421.
31. Malley DS, Steinert RF, Puliafito CA, et al: Immunofluorescence study of corneal wound healing after excimer laser anterior keratectomy in the monkey eye. *Arch Ophthalmol* 1990;108:1316–1322.
32. Hanna KD, Pouliquen Y, Waring GO: Corneal stromal healing in rabbits after 193 nm excimer laser surface ablation. *Arch Ophthalmol* 1989;107:895–901.
33. Durrie DS, Lesher MP, Cavanaugh TB: Classification of variable clinical response after photorefractive keratectomy for myopia. *J Refract Surg* 1995;11: 341–347.
34. Sher NA: Postoperative management after excimer PRK. Reshaping the future: Refractive surgery. 1995, pp 26–28.
35. Machat JJ: PRK Complications and their management. In *Excimer Laser Refractive Surgery, Practice and Principles*. Thorofare, NJ, SLACK Inc, 1996.
36. Yang H, Toda I, Bissen-Miyajima H, et al: Allergic conjunctivitis is a risk factor for regression and haze after PRK. 1995 ISRS Pre-AAO Conference and Exhibition, October 26, 1994, Atlanta, GA.
37. Maloney RK, Chan WK, Steinert R: A multicenter trial of photorefractive keratectomy for residual myopia after previous ocular surgery. *Ophthalmology* 102(7):1042–1053.
38. Seiler T, Kahle G, Kriegerowski M: Excimer laser (193nm) myopic keratomileusis in sighted and blind human eyes. *Refract Corneal Surg* 1990;6:165–173.
39. Tuft S, Zabel RW, Marshall JW: Corneal repair following keratectomy: a comparison between conventional surgery and laser photoablation. *Invest Ophthalmol Vis Sci* 1989;30:1769–1777.
40. Taylor DM, L'Esperance FA, Del Pero RA, et al: Human excimer laser lamellar keratectomy: a clinical study. *Ophthalmology* 1989;96:654–664.
41. Gartry DS, Muir MGK, Marshall J: The effect of topical corticosteroids on refraction and corneal haze following excimer laser treatment of myopia: an update. A prospective, randomized, double-masked study. *Eye* 1993;7:584–590.
42. Nassaralla BA, Szerenyi K, Wang XW, et al: Effect of diclofenac on corneal haze after photorefractive keratectomy in rabbits. *Ophthalmology* 1995;102(3): 469–474.
43. Sher NA, Frantz JM, Talley A: Topical diclofenac in the treatment of ocular pain after excimer PRK. *Refract Corneal Surg* 1993;9:425–442.
44. Jain S, Hahn TW, McCally R, et al: Antioxidants reduce corneal light scattering after excimer keratectomy in rabbits. *Lasers Surg Med* 1995;17:160–165.
45. Lohman CP, Marshall J: Plasmin- and plasminogen activator inhibitors after excimer laser photorefractive keratectomy: new concept in prevention of postoperative myopic regression and haze. *Refract Corneal Surg* 9:300–302.
46. Morlet N. Gillies MC, Crouch R: Effect of topical Alpha 2b-interferon on corneal haze after excimer laser PRK in rabbits. *Refract Corneal Surg* 1993;9: 443–451.
47. Jain S, Khoury JM, Chamon W, et al: Corneal light scattering after laser in situ keratomileusis and photorefractive keratectomy. *Am J Ophthalmol* 1995; 120:532–534.
48. Tervo T, Mustronen R, Tarkkanen A: Management of dry eye may reduce haze after excimer laser photorefractive keratectomy [Letter]. *Refract Corneal Surg* 1990;9(Suppl):306.
49. Cantera E, Cantera I, Olivieri L: Corneal topographic analysis of photorefractive keratectomy in 175 myopic eyes. *Refract Corneal Surg* 1993;9 (Suppl):S19–S22.
50. Cavanaugh TB, Durrie DS, Riedel SM, et al: Topographical analysis of the centration of excimer laser photorefractive keratectomy. *J Cataract Refract Surg* 1993;19:136–143.
51. Wilson SE, Klyce SD, McDonald MB, et al: Changes in corneal topography after excimer laser photorefractive keratectomy for myopia. *Ophthalmology* 1991;98:1338–1347.
52. Cavanaugh TB, Durrie DS, Riedel SM, et al: Centration of excimer laser photorefractive keratectomy relative to the pupil. *J Cataract Refract Surg* 1993;19:144–148.

53. Amano S, Tanaka S, Shimizu K: Topographical evaluation of centration of excimer laser myopic photorefractive keratectomy. *J Cataract Refract Surg* 1994;20:616–619.
54. Roberts C: Characterization of the inherent error in a spherically-biased corneal topography system in mapping radially aspheric surface. *J Cataract Refract Surg* 1994;10:103–116.
55. Roberts C: The accuracy of "power" maps to display curvature data in corneal topography systems. *Invest Ophthalmol Vis Sci* 1994;35(9):3525–3532.
56. Klein SA, Mandell RB: Axial and instantaneous power conversion in corneal topography. *Invest Ophthalmol Vis Sci* 1995;36:2155–2159.
57. Fay AM, Trokel SL, Myers JA: Pupil diameter and the principal ray. *J Cataract Refract Surg* 1992;18:348–351.
58. Klyce SD, McDonald MB: Computerized corneal topography of surface ablations with the Tomey (TMS-1). In Salz JJ, McDonnell PJ, McDonald MB (eds): *Corneal Laser Surgery*. St Louis, MO, Mosby, 1995, pp 100–101.
59. Klyce SD, Smolek MK: Corneal topography of the excimer photorefractive keratectomy. *J Cataract Refract Surg* 1993;19(Suppl):122–130.
60. Maguire LJ, Zabel RW, Parker P, et al: Topography and raytracting analysis of patients with excellent visual acuity 3 months after excimer laser photorefractive keratectomy for myopia. *Refract Corneal Surg* 1991;7:122–128.
61. Uozato H, Guyton DL: Centering corneal surgical procedures. *Am J Ophthalmol* 1987;103:264–275.
62. Seiler T, Reckmann W, Maloney RK: Effective spherical aberration of the cornea as a quantitative descriptor in corneal topography. *J Cataract Refract Surg* 1993;19(Suppl):155–165.
63. Azar DT, Yeh P: Corneal topographical evaluation of decentration in PRK: Treatment displacement (shift) vs. drift. *Am J Ophthalmol* 1997 in press.

section five

FDA-Sponsored PTK Clinical Trials

CHAPTER 13
PTK Results: Summit Excimer Laser

Roger F. Steinert

This chapter will present a detailed summary of the clinical results achieved with the Summit Technology excimer laser for phototherapeutic keratectomy and a brief summary of the results of photorefractive keratectomy. Early experience utilized the Summit Technology Excimer UV 200 system; more recent experience has been with the Summit Technology Apex excimer laser (originally termed Omnimed). The major difference in those two systems was the expansion of the optical zone to 6.0 mm in the Apex system compared to 5.0 mm in the Excimer system. The output beam characteristics of the two systems are otherwise physically indistinguishable.

Clinical investigators of the phototherapeutic keratectomy studies are listed in Table 13.1.

PHOTOTHERAPEUTIC KERATECTOMY

These results summarize the presentation of data made to the Ophthalmic Advisory Panel of the U.S. Food and Drug Administration on March 21, 1994. Phototherapeutic keratectomy (PTK) with the Summit Technology laser was approved and premarket approval granted for the Summit Technology system on March 10, 1995. The PTK procedures were performed utilizing a standardized protocol throughout all phases of the study and at all centers, allowing pooling of the multicenter data. All PTK procedures were therapeutic only, with no refractive component.

A total of 398 eyes were enrolled and results reported under the PTK protocol. The preoperative pathology was divided into five categories: (1) anterior stromal dystrophies; (2) epithelial basement membrane dystrophy; (3) recurrent erosions; (4) other surface irregularity; and (5) anterior scars. This classification and the treatment goal, as well as the number of eyes enrolled in each category, are listed in Table 13.2.

One of three clinical techniques was employed, depending upon the preoperative pathology:

- *Polish technique.* This technique was recommended for relatively smooth corneal surfaces where superficial opacity was the principal pathology. Ablation could be performed through the epithelium or after mechanical deepithelialization. Judicious use of artificial tear drops as a masking agent was usually employed with a large spot size. The patient's head was slightly moved in an orbital rotation to "blend" the edges of the ablation with the untreated peripheral cornea.

- *Debride and polish technique.* This technique was recommended for highly irregular surfaces. Opacities as well as irregularity might be contributing factors to the visual impairment. In most cases, ablation was performed through the epithelium with a large spot size, judging penetration of the epithelium by the disappearance of the normal epithelial fluorescence, which is visible under dim illumination. After ablation through the epithelium, which itself acts as a masking agent, the surface was mechanically débrided with a wiping motion utilizing a rounded ophthalmic microsurgical blade. Further ablation was then performed, with frequent applications of an articial tear solution as a masking agent. In cases of highly localized irregularity or opacity, an appropriately smaller spot size was employed, with the beam directed specifically at the pathology. A larger beam was then used to create a smoother surface blending the area of pathology into the noninvolved cornea. (See Chapter 8)

Table 13.1. Investigational Center/Principal Investigator List for IDE G880234 PTK

Investigational Center	Principal Investigator(s)	Coinvestigator(s)
1. Cleveland Clinic Foundation Cleveland, OH	Roger Langston, M.D. Gary Varley, M.D. (prev.)	David M. Meisler, M.D.
2. Emory University Atlanta, GA	George Waring, M.D.	R. Doyle Stulting, M.D.
3. Eye Physicians of Omaha Omaha, NE	Jeffery Hottmann, M.D.	
4. Georgetown University Washington, DC	Jay Lustbader, M.D. Rajesh Rajpal, M.D. (prev.)	Jonathan Javitt, M.D.
5. Grove Hill Medical Center New Britain, CT	Alan Stern, M.D.	Daniel M. Taylor, M.D.
6. Hunkeler Eye Clinic Kansas City, MO	Daniel Durrie, M.D.	Tim Cavanaugh, M.D. John Hunkeler, M.D.
7. John Eye Clinic, PSC Jeffersonville, IN	Maurice John, M.D.	
8. Jones Eye Clinic Sioux City, IA	Charles Jones, M.D.	
9. Jules Stein Eye Institute Los Angeles, CA	Robert Maloney, M.D.	Richard Elander, M.D.
10. Mann/Berkeley Eye Clinic Houston, TX	Ralph Berkeley, M.D. Stephen Slade, M.D. (prev)	Michael Mann, M.D.
11. Mansfield Professional Bldg. Burlington, VT	David Chase, M.D.	Michael Vrabec, M.D.
12. Montefiore Hospital Bronx, NY	Lewis Grodin, M.D. Peter Hersh, M.D. (prev.)	Martin Mayers, M.D.
13. Ophthalmic Consultants of Boston Boston, MA	Roger Steinert, M.D.	Carmen Puliafito, M.D. Mariana Mead, M.D. Michael Raizman, M.D. Helen Wu, M.D.
14. Pacific Cataract and Laser Institute Chehalis, WA	Robert Ford, M.D. Helgi Heidar, M.D. (prev.)	
15. Sioux Empire Medical Center Sioux Falls, SD	Vance Thompson, M.D.	Byron Hohm, M.D.
16. Vision Surgery and Laser Center San Diego, CA	Michael Gordon, M.D.	C.L. Blanton, M.D. Steve Schallhorn, M.D.
17. Washington University Saint Louis, MO	Jay Pepose, M.D., Ph.D.	Larry Gans, M.D. (prev.)
18. Western Pennsylvania Eye Center Pittsburgh, PA	Edward Kondrot, M.D.	

Table 13.2. Patient Treatment Categories

1. Anterior Stromal Dystrophies — 38 eyes* (9.6%)
   (alternative generally PK)
   Goal: usually to improve vision
   Including: Granular
   Lattice
   Reis-Bücklers'
   Macular
2. Epithelial Basement — 65 eyes (16.3%)
   Membrane Dystrophy
   (alternative generally *not* PK/LK)
   Goal: to improve vision or comfort
   Cogan's map-dot-fingerprint dystrophy
3. Recurrent Erosions — 64 eyes (16.1%)
   Goal: to improve comfort
   Including: Traumatic
   Nontraumatic
   Postinfectious
4. Other Irregular Surfaces — 116 eyes (29.1%)
   Goal: to improve vision and/or comfort
   Including: Salzmann's degeneration
   Spheroid degeneration
   Postsurgical scars
   Keratoconus nodules
5. Scars — 115 eyes (28.9%)
   Goal: usually to improve vision
   Including: After trauma
   After infection
   Pathologic

*A total of 398 eyes were enrolled and reported here.
PK, penetrating keratoplasty; LK, lamellar keratoplasty.

Table 13.3. PTK Technique Issues

- Recommended clinical techniques
- Polish technique (recommended for relatively smooth corneal surfaces)
- Debride and polish technique (recommended for highly irregular surfaces)
- Recurrent erosion technique (15 to 40 laser pulses with polish technique)

No statistically significant difference was found in success rates at 3 ($p = 0.204$), 6 ($p = 0.194$) and 12 ($p = 0.639$) months based on the clinical technique utilized

- The PRK refractive algorithm was *not* utilized in the Summit PTK clinical investigation

Clinical technqiue *not* recommended

- Point and shoot technique

- *Recurrent erosion technique.* For recurrent erosions, either traumatic or associated with anterior basement membrane dystrophy, the epithelium was first debrided mechanically, either with the blade or with a dry cellulose surgical spear. The area of poor epithelial adhesion was thereby defined. This area often was larger in extent than the area of breakdown documented in preoperative exams, consistent with widespread anterior basement membrane dysfunction. The involved area was then treated with 15 to 40 laser pulses. The most common approach was to use a minimal number of pulses, such as 16, to remove an average of only 4 μm of tissue. If the area of pathology exceeded the maximum beam diameter, slight overlap of multiple treatment zones occurred. Movement of the patient's head in a "polish" fashion helped avoid abrupt transitions of treated, nontreated, and repeatedly treated areas. (See Chapter 9)

The photorefractive keratectomy (PRK) refractive algorithm was not utilized for any PTK treatment. The *point and shoot* technique of simply allowing the laser to fire at opacities or irregularities without the use of masking agents or polishing techniques, was not recommended or employed.

The reported follow-up interval for each of the treated eyes is given in Table 13.4. At the time of data analysis, slightly more than 50% of the eyes had at least 1 year of follow-up.

After PTK, most eyes reepithelialize within 1 week. Delays in reepithelialization were attributable to the investigational nature of this technique, with treatments being performed on patients with severe

Table 13.4. Long-Term Follow-up Periods

3 months	333 eyes
6 months	276 eyes
12 months	206 eyes
24 months	94 eyes
36 months	29 eyes

Table 13.5. Reepithelialization

Following PTK, eyes reepithelialized by the following times:

By 1 day	13 eyes	(3.3%)
By 3 days	245 eyes	(61.6%)
By 1 week	345 eyes	(86.7%)
By 1 month	387 eyes	(97.2%)
By 3 months	391 eyes	(98.2%)

Reepithelialization time is unknown in the remaining 7 eyes.

In the 4 eyes that took 1 to 3 months to reepithelialaize, the following underlying diseases were present:
- Preexisting bullous keratopathy and recurrent lattice dystrophy
- Preexisting rosacea keratitis, ocular chlamydia, and follicular conjunctivitis which caused an irregular tear film
- Preexisting neurotrophic keratitis, exposure keratitis, and seventh nerve palsy
- Secondary to toxic effect of antibiotics and underlying chronic disease

underlying disease frequently associated with healing defects (Table 13.5).

Best spectacle corrected visual acuity was statistically significantly improved at each of the principal follow-up intervals throughout the first postoperative year. As shown in Table 13.6, the proportion of patients with 20/40 or better best spectacle corrected visual acuity increased by nearly 22%, while the proportion with 20/100 or worse acuity decreased by slightly over 10% at 1 year. Also notable is the stability of the visual acuity levels at 3, 6, and 12 months. The improvement in visual acuity is more striking when a subgroup analysis is performed of the patients in whom the therapeutic goal was to improve vision (Table 13.7). Best corrected visual acuity of 20/40 or better increased by over 33% and acuity of 20/100 or worse decreased by nearly 16% at 1 year. Again the proportion of patients in each visual acuity category is stable at 3, 6, and 12 months of follow-up. When glare is simulated with the brightness acuity tester (B.A.T.) (Mentor, Inc., Norwell, MA) set at medium, further degradation of vision to levels below 20/100 is seen preoperatively; a decrease of nearly 40% in the proportion of patients with acuity of 20/100 or worse is seen after PTK (Table 13.8).

Table 13.6. Best Corrected Visual Acuity (BCVA)

	Preop ($n = 387$)	3 Months ($n = 311$)	6 Months $n = 262$	1 Year ($n = 201$)
20/40 or better	176 (45.5%)	187 (60.1%)	169 (65.5%)	135 (67.1%)
20/50–20/80	117 (30.2%)	69 (22.2%)	50 (19.1%)	38 (18.9%)
20/100–20/400	67 (17.3%)	33 (10.6%)	25 (9.5%)	14 (7.0%)
Less than 20/400	27 (7.0%)	22 (7.1%)	18 (6.9%)	14 (7.0%)

- BCVA of 20/40 or better increased by 21.6% at 1 year.
- BCVA of 20/100 or worse decreased by 10.3% at 1 year.

The distribution of acuities is significantly different at 3.6 and 12 months compared with the preoperative status using the chi-square test performed with SAS system 6.07.

Distribution of acuities preoperatively compared with distribution of acuities at 3 months: χ^2 with 3 degrees of freedom = 16.714 ($p < .001$).

Distribution of acuities preoperatively compared with distribution of acuities at 6 months: χ^2 with 3 degrees of freedom = 24.842 ($p < .001$).

Distribution of acuities preopeatively compared with distribution of acuities at 12 months: χ^2 with 3 degrees of freedom = 28.484 ($p < .001$).

Table 13.7. Best Corrected Visual Acuity by Goal

Goal: Improve Vision

	Preop (n = 228)	3 months (n = 192)	6 months (n = 164)	1 year (n = 133)
20/40 or better	67 (29.4%)	100 (52.1%)	95 (57.9%)	84 (63.1%)
20/50–20/80	96 (42.1%)	61 (31.8%)	44 (26.8%)	32 (24.1%)
20/100–20/400	55 (24.1%)	24 (12.5%)	19 (11.6%)	12 (9.0%)
Less than 20/400	10 (4.4%)	7 (3.6%)	6 (3.7%)	5 (3.8%)

- BCVA of 20/40 or better increased by 33.7% at 1 year.
- BCVA of 20/100 or worse decreased by 15.7% at 1 year.

Goal: Improve comfort

	Preop (n = 154)	3 Months (n = 115)	6 Months (n = 94)	1 Year (n = 66)
20/40 or better	106 (68.9%)	84 (73.0%)	71 (75.5%)	49 (74.3%)
20/50–20/80	21 (13.6%)	8 (7.0%)	6 (6.4%)	6 (9.1%)
20/100–20/400	11 (7.1%)	8 (7.0%)	5 (5.3%)	2 (3.0%)
Less than 20/400	16 (10.4%)	15 (13.0%)	12 (12.8%)	9 (13.6%)

- BCVA of 20/40 or better increased by 5.4% at 1 year.
- BCVA of 20/100 or worse decreased by 0.9% at 1 year.

Table 13.8. Best Corrected Visual Acuity with Glare by Goal

Goal: improve vision

	Preop (n = 170)	3 Months (n = 123)	6 Months (n = 105)	1 Year (n = 98)
20/40 or better	22 (12.9%)	55 (44.4%)	47 (44.8%)	47 (47.9%)
20/50–20/80	59 (34.7%)	37 (29.8%)	33 (31.4%)	38 (38.8%)
20/100–20/400	72 (42.4%)	25 (21.0%)	21 (20.0%)	10 (10.2%)
Less than 20/400	17 (10.0%)	6 (4.8%)	4 (3.8%)	3 (3.1%)

- BCVA w/glare of 20/40 or better increased by 35.0% at 1 year.
- BCVA w/glare of 20/100 or worse decreased by 39.1% at 1 year.

Goal: improve comfort

	Preop (n = 121)	3 Months (n = 85)	6 Months (n = 68)	1 Year (n = 45)
20/40 or better	70 (57.8%)	56 (65.9%)	49 (72.0%)	29 (64.4%)
20/50–20/80	19 (15.7%)	11 (12.9%)	6 (8.8%)	8 (17.8%)
20/100–20/400	22 (18.2%)	9 (10.6%)	5 (7.4%)	3 (6.7%)
Less than 20/400	10 (8.3%)	9 (10.6%)	8 (11.8%)	5 (11.1%)

- BCVA w/glare of 20/40 or better increased by 6.6% at 1 year.
- BCVAa w/glare of 20/100 or worse decreased by 8.7% at 1 year.

Although PTK is performed with a planar disc pattern, which should not induce any immediate refractive shift, the postoperative healing of the cornea may result in a shift in corneal optics. The principal reasons for a myopic correction (hyperopic shift) are epithelial hyperplasia, and stromal healing with new collagen deposition greater at the periphery of the ablation zone and less toward the center. An overall flattening then results. Hyperopic correction (myopic shift) occurs most commonly when the pathology undergoing ablation is eccentric, so that more midperipheral corneal tissue than central corneal tissue is removed, resulting in an overall steepening of the optical zone. Table 13.9 details the refractive status of the 159 eyes, with the refraction available both preoperatively and 12 months after PTK. Nearly half of the eyes remained within ± 1 D of the preoperative refraction. Of those with a refractive change greater than 1 D, however, a hyperopic shift was much more common than a myopic shift (38.4% versus 15.1%). Figure 13.1 shows the mean spherical equivalent over time. The refractive status appears to be stable at 2 and 3 years postoperatively compared to the 1-year refraction.

Patients were classified as a success, a failure, or no change. Where the treatment goal was to improve vision, success was defined as gain of more than two lines of visual acuity and failure as loss of more than two lines of visual acuity. Where the goal was to improve comfort, success was defined as patient satisfaction that comfort had been restored a substantial proportion of the time compared to the preoperative situation. Tables 13.10 to 13.14 give the outcomes at 3, 6, 12, 24, and 36 months postoperatively. Figures 13.2 to 13.4 graphically depict the percentage of patients classified as successful over time. For PTK treatments as a whole, the proportion judged to be successful is approximately 75% at all postoperative intervals. Where the goal was to improve comfort, the success rate is approximately 85%; where the goal was to improve

Table 13.9. Refractive Change Following PTK, Preoperatively to 12 Months (N = 159)

The 159 eyes with a refraction available both preoperatively and at 12 months had the following changes in manifest refraction spherical equivalent:

- 46.5% of eyes were within ± 1 D of the preoperative value
- 66.7% of eyes were within ± 2 D of the preoperative value
- 77.4% of eyes were within ± 3 D of the preoperative value
- 15.1% of the eyes shifted toward myopia
- 38.4% of the eyes shifted toward hyperopia
- 46.5% of the eyes were unchanged

Specifically:

6.25 to 7.00 D toward myopia	1 eye	0.6%
5.25 to 6.00 D toward myopia	1 eye	0.6%
4.25 to 5.00 D toward myopia	4 eyes	2.5%
3.25 to 4.00 D toward myopia	3 eyes	1.9%
2.25 to 3.00 D toward myopia	3 eyes	1.9%
1.25 to 2.00 D toward myopia	12 eyes	7.6%
No change (within ± 1.00 D)	74 eyes	46.5%
1.25 to 2.00 D toward hyperopia	20 eyes	12.6%
2.25 to 3.00 D toward hyperopia	14 eyes	8.8%
3.25 to 4.00 D toward hyperopia	14 eyes	8.8%
4.25 to 5.00 D toward hyperopia	1 eye	0.6%
5.25 to 6.00 D toward hyperopia	3 eyes	1.9%
6.25 to 7.00 D toward hyperopia	6 eyes	3.8%
7.25 to 8.00 D toward hyperopia	0 eyes	0.0%
8.25 to 9.00 D toward hyperopia	3 eyes	1.9%

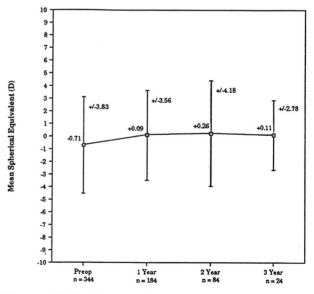

Figure 13.1. Mean spherical equivalent in PTK patients over time. Error bars represent ± 1 standard deviation.

Table 13.10. Success Rates 3 Months Postoperatively

Overall results (N = 333)		
Success	243 eyes	73.0%
No change	35 eyes	10.5%
Failure	55 eyes	16.5%
Goal: improve vision (N = 201)		
Success	127 eyes	63.2%
No change	35 eyes	17.4%
Failure	39 eyes	19.4%
Goal: improve comfort (N = 127)		
Success	111 eyes	87.4%
Failure	16 eyes	12.6%

Table 13.11. Success Rates 6 Months Postoperatively

Overall results (N = 276)		
Success	206 eyes	74.6%
No change	24 eyes	8.7%
Failure	46 eyes	16.7%
Goal: improve vision (N = 167)		
Success	112 eyes	67.0%
No change	23 eyes	13.8%
Failure	32 eyes	19.2%
Goal: improve comfort (N = 104)		
Success	90 eyes	86.5%
Failure	14 eyes	13.5%

Table 13.12. Success Rates 12 Months Postoperatively

Overall results (N = 206)		
Success	151 eyes	73.3%
No change	23 eyes	11.2%
Failure	32 eyes	15.5%
Goal: improve vision (N = 136)		
Success	88 eyes	64.7%
No change	23 eyes	16.9%
Failure	25 eyes	18.4%
Goal: improve comfort (N = 127)		
Success	61 eyes	89.7%
Failure	7 eyes	10.3%

Table 13.13. Success Rates 2 Years Postoperatively

Overall results (N = 94)		
Success	68 eyes	72.3%
No change	14 eyes	14.9%
Failure	12 eyes	12.8%
Goal: improve vision (N = 61)		
Success	40 eyes	65.5%
No change	14 eyes	23.0%
Failure	7 eyes	11.5%
Goal: improve comfort (N = 32)		
Success	27 eyes	84.4%
Failure	5 eyes	15.6%

Table 13.14. Success Rates 3 Years Postoperatively

Overall results (N = 29)		
Success	20 eyes	69.0%
No change	3 eyes	10.3%
Failure	6 eyes	20.7%
Goal: improve vision (N = 15)		
Success	9 eyes	60.0%
No change	3 eyes	20.0%
Failure	3 eyes	20.0%
Goal: improve comfort (N = 12)		
Success	9 eyes	75.0%
Failure	3 eyes	25.0%

vision, the success rate is approximately 66%. Of particular note is the stability of the proportion of patients judged to be a success or failure over time. By 3 months after PTK, success or failure is usually evident.

Patient questionnaires were completed preoperatively and at 3, 6, and 12 months postoperatively. Tables 13.15 and 13.16 give the key results. Notably, approximately 85% of patients were satisfied enough with their results that they would agree to have PTK again.

The principal advantages of PTK compared to alternative forms of corneal surgery are listed in Table 13.17. PTK is a more artful procedure than PRK; with increased experience, the ophthalmic

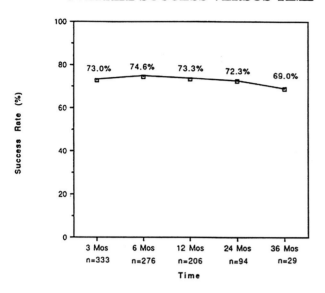

Figure 13.2. Success rate over time for PTK patients as a whole.

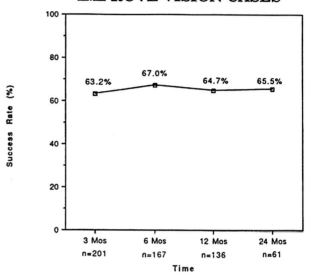

Figure 13.4. Success rate over time for patients undergoing PTK with the goal of improving vision.

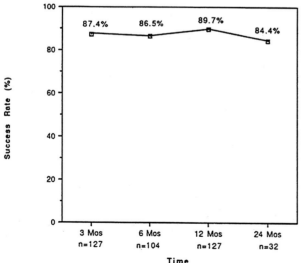

Figure 13.3. Success rate over time for patients undergoing PTK with the goal of improving comfort.

surgeon will experience higher success rates than those reported here by investigators with no prior experience, no body of knowledge from experienced predecessors, and no established guidelines. Through proper case selection, judicious and skillful application of laser energy in conjunction with the use of masking fluids and mechanical débridement, and appropriate postoperative management, a surgeon's individual results rapidly improve.

The appropriate treatment goals differ by type of preoperative pathology. In the case of anterior basement membrane dystrophy with disruption of vision and/or discomfort, PTK may well be curative. In the case of anterior stromal dystrophies, which are recurrent, the goal of PTK is to improve vision to an acceptable functional level but rarely to normalize vision entirely. By removing the most recently developing opacities, which tend to be the most superficial, PTK may delay the need for keratoplasty and thereby reduce the total number of keratoplasties necessary over the lifetime of the patient.

Table 13.15. Patient Survey Results

Number of completed patient surveys
Preoperative   316 surveys
3 months   195 surveys
6 months   174 surveys
1 year   124 surveys

- Would you have PTK again? (YES)
  3 months   83.6%
  6 months   87.4%
  1 year   85.5%

Goal: improve vision

- Percent of patients *very satisfied* with vision (4 or 5 on 0–5 scale)
  Preoperative ($n = 158$)   10.8%
  1 year ($n = 87$)   59.6%

- Percent of patients *very dissatisfied* with vision (0 or 1 on 0–5 scale)
  Preoperative ($n = 158$)   35.5%
  1 year ($n = 87$)   7.0%

Goal: improve comfort

- Percent of patients reporting *a great deal of pain* in eye
  Preoperative ($n = 137$)   53.2%
  1 year ($n = 37$)   2.7%

- Percent of patients reporting *no pain at all* in eye
  Preoperative ($n = 137$)   8.0%
  1 year ($n = 37$)   48.7%

Table 13.16. Patient Survey Results

Percentage of patients reporting severe problems (4 or 5 on 0–5 scale) with the following decreased after PTK:

Glare	Preop: 34.6%	1 year: 18.6%
Halo	Preop: 25.6%	1 year: 9.6%
Problems seeing in dark	Preop: 29.4%	1 year: 14.5%
Difficulty reading	Preop: 41.6%	1 year: 20.2%
Difficulty with distance vision	Preop: 42.7%	1 year: 16.2%
Distortion	Preop: 29.8%	1 year: 8.8%
Blurring	Preop: 36.0%	1 year: 15.4%
Problems seeing in bright lights	Preop: 44.2%	1 year: 25.8%
Eye strain	Preop: 27.1%	1 year: 12.9%
Pain	Preop: 26.3%	1 year: 6.4%
Tearing	Preop: 25.4%	1 year: 8.0%
Halo	Preop: 25.6%	1 year: 9.6%
Light sensitivity	Preop: 43.1%	1 year: 17.0%
Redness	Preop: 20.2%	1 year: 3.2%
Foreign body sensation	Preop: 25.9%	1 year: 8.9%

Table 13.17. Alternate Practices/Procedures

Key advantages of PTK

- A minimally invasive procedure
- Performed under topical anesthesia
- Office or hospital setting (outpatient basis)
- The 193-nm laser allows precise tissue removal with minimal surrounding tissue damage, leaving a smooth optical surface *unsurpassed* by diamond/metal knives
- Does *not* preclude performance of more invasive technique (i.e., penetrating keratoplasty)
- An extraocular procedure: superficial cornea
- Risk of infection is *low*
- Does *not* weaken overall corneal integrity
- Visual recovery/convalescence time is short
- *Not* dependent on donor material

Although the number of candidates for PTK constitutes a small percentage of the patients in a general ophthalmic practice, PTK is an important new treatment modality that offers an attractive therapeutic alternative. For the ophthalmic surgeon, whether a comprehensive general ophthalmologist or a corneal specialist, the excimer laser is a major addition to the therapeutic armamentarium.

CONCLUSION

The excimer laser operating at 193-nm is a powerful new tool in corneal therapeutics. The principal interest in the excimer laser has been for treatment of refractive error. The current results justify the early enthusiasm for this technology.

PTK has been a fortuitous benefit of the development fueled by the refractive applications of the excimer laser. Although PTK will not eliminate corneal transplantation, it does provide a therapeutic alternative for the treatment of superficial corneal opacities and irregularities that delays and perhaps eliminates the need for lamellar or penetrating keratoplasty in favorable patients.

CHAPTER 14
PTK Results: VISX Excimer Laser

Farooq Ashraf, Dimitri T. Azar, Marc Odrich

The VISX excimer laser has been approved by the Food and Drug Administration (FDA) for both refractive and therapeutic uses. As discussed in previous chapters, phototherapeutic keratectomy (PTK) creates a superficial lamellar keratectomy by removing the anterior layers of the cornea with submicron precision without significant injury to adjacent nonablated tissue. This chapter summarizes the results of the VISX-sponsored prospective, nonrandomized, uncontrolled, unmasked multicenter clinical study that was approved under an investigational device exemption granted from the FDA to determine the safety and efficacy of PTK for therapeutic purposes.

A total of 269 primary eyes were treated in 17 clinical centers (Table 14.1). This study included the examination protocol for the preoperative, perioperative, intraoperative, and immediate as well as long-term, postoperative periods. Patients were greater than 18 years of age, had best corrected visual acuity worse than 20/40, had corneal pathology in the anterior third of the cornea, and the immediate postoperative corneal thickness was expected to be at least 250 µm.

Indications for treatment included patients with decreased best corrected visual acuity from anterior stromal scars or opacities and/or disabling pain from epithelial irregularities in the cornea that have failed alternative nonsurgical treatment options (Table 14.2). Contraindications to treatment were scars not contained in the anterior third of the cornea, potential postoperative thickness of less than 250 µm, individuals who could not tolerate the procedure, the inability to comply with postoperative instructions, patients with significant ocular surface disease such as severe blepharitis, lagophthalmos, dry eyes, active uveitis, or individuals who were immunocompromised (Table 14.2). Success was defined as either an improvement in vision or comfort.

PREOPERATIVE EXAMINATION

Preoperative examinations were performed within 90 days of treatment. Visual acuity with and without correction was measured along with visual potential by using pinhole, hard contact lens, or potential acuity meter. A careful anterior segment examination was performed to determine the extent of corneal pathology by slit-lamp examination. Corneal thickness and extent of corneal pathology were measured by an optical pachymeter. Best spectacle corrected visual acuity (BSCVA) showed 18% 20/40 or better, 42% between 20/50–20/100, and 42% worse than 20/100 (Table 14.3). A four-point scale was used to assess patient symptoms of pain, tearing, photophobia, conjunctival erythema, or foreign body sensation; and of these only the moderate to severe categories were included in the study. Preoperative corneal clarity was scored as: none–mild if refraction was not affected, moderate if refraction was affected but not prevented, and severe if unable to refract. Patients with a history of herpes simplex were prophylaxed with oral acyclovir 1,000 mg a day starting 2 days prior to treatment for a total of 14 days. Topical Viroptic was matched drop for drop with the topical steroids for a total of 2 weeks postoperatively.

PERIOPERATIVE PREPARATION

Clinical experience has shown that PTK is well tolerated and rarely causes significant pain. Topical anesthesia, such as 0.5% proparacaine drops, was

Table 14.1. Patient Accountability

n = 269
240 examined in 3 months
146 at 6 months
138 at 1 year
222 ≥ 1 year
3% re-treated, 3% died, 10% lost to follow-up

Table 14.2.

Inclusion Criteria	Exclusion Criteria
Age ≥ 18 yr	Immunosuppressed individual
BSCVA ≤ 20/40	Uncontrolled uveitis
Pathology in anterior one-third of the cornea	Severe blepharitis
Postoperative corneal thickness ≥ 250 μm	Lagophthalmos
	Severe dry eyes
	Inability to tolerate the procedure, to lie flat, or inability to comply with postoperative therapy

BSCVA = best spectacle corrected visual acuity.

Table 14.3. Preoperative Characteristics

	BSCVA (%)	UCVA (%)
20/40 or better	18	2
20/50–20/100	42	21
20/100 or worse	40	77

BSCVA = best spectacle corrected visual acuity;
UCVA = uncorrected visual acuity.

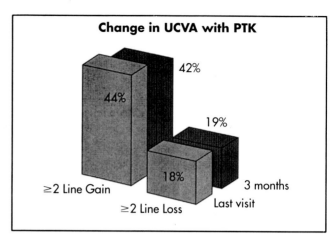

Figure 14.1. Change in uncorrected visual acuity with PTK. Improvement in uncorrected visual acuity, while not a defining goal, was achieved. From 42% to 44% of patients had a gain of two or more lines of uncorrected acuity, and only 18% to 19% had an equivalent loss. (From Azar DT et al: Phototherapeutic keratectomy: the VISX experience. In Salz JJ, McDonnell PJ, McDonald MB, editors: *Corneal laser surgery*, St Louis, 1995, Mosby.)

usually the only anesthetic needed for the procedure. Systemic sedatives, such as diazepam, were given to excessively anxious patients.

INTRAOPERATIVE TECHNIQUE

Prior to each treatment the laser was successfully calibrated. Table 14.4 summarizes the laser parameters that are ordinarily adopted for PTK using the VISX laser. The epithelium is removed mechanically or by the laser. This decision is based on the smoothness of the epithelium relative to the envisioned smoothness of Bowman's layer. When the anterior stromal surface is irregular, the epithelium acts as a smoothing agent and is ablated with the laser. If the anterior surface of the stroma is judged to be smooth, the epithelium may be removed manually with a surgical blade. In 265 of 269 eyes where the epithelial removal technique was specified, 56% were manually removed and 44% were removed by the laser. Following epithelial removal, a masking fluid or smoothing agent such as artificial tears of methylcellulose may be applied to help achieve a smooth stromal surface for ablation. Ablation of

Table 14.4. Typical Laser Parameters for PTK Using VISX Excimer Laser

Fluence	160 ± 10 mJ/cm^2
Repetition rate	5 Hz
Ablation rate	0.20–0.35 μm/pulse
Ablation diameter	5.5–6.0 mm including 0.5 mm transition zone
Ablation depths:	
Epithelium	40 μm or as determined by pachymetry
Stroma	Depth of scar or opacity

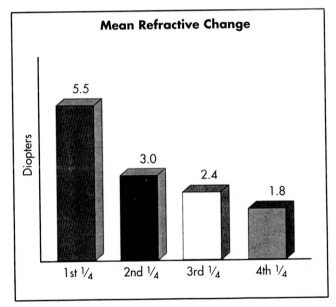

Figure 14.2. Mean refractive change. Patients were divided into four groups or quartiles based on the date of treatment. The group or quartile that was treated initially had a change in spherical equivalent of 5.5 diopters. By the second quartile the mean had been decreased by almost 50% to 3.2 diopters. The last series demonstrated less than 2 diopters of refractive effect. Thus it appears that marked hyperopic shifts are no longer a problem. (From Azar DT et al: Phototherapeutic keratectomy: the VISX experience. In Salz JJ, McDonnell PJ, McDonald MB, editors: *Corneal laser surgery*, St Louis, 1995, Mosby.)

corneal pathology is performed until the desired tissue is removed. If any uncertainty occurs, a quick examination using a slit lamp is warranted. During stromal ablation a transition zone and a modified taper technique are used to promote wound healing and to decrease central flattening.

POSTOPERATIVE TECHNIQUE

Following excimer laser surgery, antibiotics, cylcoplegics, steroids, and a firm patch are applied to the eye. The patient is examined daily until the epithelium heals and the topical medications are tapered accordingly.

RESULTS

The epithelium was completely healed by 4 days in 70% of cases and in 97% by 30 days. Only in 1% (4 eyes) did the epithelium fail to completely heal. These patients had herpetic scars, neurotrophic keratopathy, and two patients had alkali burns. At 1 year postoperatively, 53% gained 2 or more lines of BSCVA, while 8% lost two or more lines. 41% gained 3 or more lines of BSCVA, while only 6.8% lost two or more (Table 14.5). The average refractive change was 2.4 and 2.3 diopter hyperopic shift at 3 months and 1 or more years post-treatment respectively (Table 14.6). 32% were ≥3.0 D hyperopic at 1 year compared to only 8% preoperatively (Table 14.6). Conversely 19% were ≥3.0 D myopic pretreatment compared to only 9% post-treatment at 1 year.

Corneal clarity of the superficial epithelium, deep epithelium, and the anterior stroma had improved. A highly significant ($p < 0.0001$) improvement in mean clinical clarity for all three layers of the cornea occurred. The improvement in visual and refractive parameters is reflected in the remarkable improve-

Table 14.5. Visual Efficacy

BSCVA gain ≥ 2 lines	53%
BSCVA gain ≥ 3 lines	41%
BSCVA lost ≥ 2 lines	8%
BSCVA lost ≥ 3 lines	6.8%

BSCVA = best spectacle corrected visual acuity.

Table 14.6. Refractive Change

3 months	+2.4 D average refractive change	Pretreatment	8% hyperopic (≥3 D)
1 yr	+2.3 D average refractive change	≥1 yr	32% hyperopic (≥3 D)

ment in corneal clarity. After PTK, patients believed that they were subjectively more comfortable (Table 14.7).

COMPLICATIONS

The most frequently encountered complications in the postoperative period were: 32% had a hyperopic shift of at least 3 D, 31% with anisometropia ≥3.0 D at 1 year, 9% with delayed reepithelialization greater than 1 week postoperatively, 8% lost 2 lines or more of BSCVA at 1 year, 3% increased corneal haze in the superficial epithelium, 5% in the deep epithelium, and 4% in the superficial stroma (Table 14.8).

Table 14.7. Improvement in Patient Symptoms

Pain	20/20
Tearing	16/17
Photophobia	49/60
Conjunctival erythema	18/20
Foreign body sensation	24/27

Table 14.8. Complications

Hyperopic shift (≥3 D)	32%
Induced anisometropia (≥3 D)	31%
Delayed reepithelialization (≥1 wk)	9%
Loss of ≥ 2 lines BSCVA (≥1 yr)	8%
Increased corneal haze	3–5%
Recurrence/reactivation of herpes	1%
Recurrent corneal dystrophy	1%
Nonherpetic corneal ulcer/infection	<1%
Corneal neovascularization	<1%

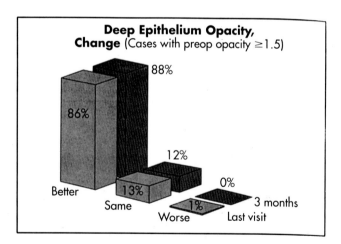

Figure 14.3. Pretreatment and posttreatment changes in deep epithelial opacity scores. From 86% to 88% of patients improved after treatment, and only 0% to 1% were worse. The patients improved a statistically significant average of 3.3 categories. (From Azar DT et al: Phototherapeutic keratectomy: the VISX experience. In Salz JJ, McDonnell PJ, McDonald MB, editors: *Corneal laser surgery*, St Louis, 1995, Mosby.)

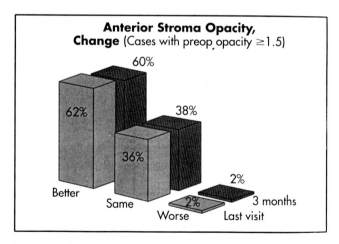

Figure 14.4. Anterior stromal opacity change. From 60% to 62% of patients improved after treatment, and 2% were worse. The patients improved a statistically significant average of 2.0 categories. (From Azar DT et al: Phototherapeutic keratectomy: the VISX experience. In Salz JJ, McDonnell PJ, McDonald MB, editors: *Corneal laser surgery*, St Louis, 1995, Mosby.)

CONCLUSION

The favorable improvement in visual and refractive parameters using the VISX excimer laser resulted directly from the substantial improvement in corneal clarity and corneal surface smoothing following PTK. This overall improvement in corneal clarity associated with corneal wound healing allowed a good return of optical clarity and a significant improvement in the level of ocular comfort for these patients. Despite these beneficial outcomes, PTK is not a panacea. Successful outcomes using the excimer laser for PTK are very much dependent on realistic expectations, good patient selection, and on the training, experience, planning, and meticulous attention to detail by the treating physician. With this in mind, PTK has the potential to benefit many patients whose only alternative treatment involves corneal surgical procedures.

REFERENCES

1. *VISX Professional Use Information Manual.*
2. Azar DT, Jain S, Woods K, et al: Phototherapeutic keratectomy: the VISX experience. In Salz JJ, McDonnell PJ, McDonald MB, editors: *Corneal Laser Surgery*, St Louis 1995, Mosby.
3. Chamon W, Azar DT, Stark WJ, et al: Phototherapeutic keratectomy. *Ophthalmol Clin North Am* 1993; 6:399–418.
4. Stark WJ, Chamon W, Kamp MT, et al: Clinical follow-up of 193-nm ArF excimer laser photokeratectomy. *Ophthalmology* 1992;99:805–812.

Index

References in italics denote figures; those followed by "t" denote tables

Ablation, (see also Photoablation)
　depth
　　description of, 9–10
　　hyperopia and, correlation between, 146
　excimer lasers for
　　calibrations, 10
　　depth, 9–10
　　mechanism of action, 8–9
　　193-nm vs. 248-nm, 10
　　threshold, 9
　laser, 100
　manual debridement, 100
　rate variations among tissues, 143–144
　refractive changes and, 143–144
　threshold levels, 9
Absorption, 24
Adhesion complex, 21
Alkali burns, to cornea, 37
Anchoring fibrils, 38
Anterior corneal dystrophies, (see Corneal dystrophies, photothereapeutic keratectomy of)
Anterior crocodile mosaic shagreen of Vogt, 74
Anterior membrane dystrophy, (see Epithelial basement membrane dystrophy)
Anterior membrane dystrophy of Grayson-Wilbrandt, 77–78
Antiviral agents, 150
Aperture wheels
　astigmatism, 9
　for hyperopia correction, 9
　for myopia correction, 8
Apex, 52
Arachidonic acid, 149
Argon fluoride gas (ArF)
　advantages of, 15
　population inversion, 6
　wavelength of, 6
Astigmatism
　against-the-rule, 56
　aperture wheels, 9
　corneal topography evaluations, 55–56
　excimer laser applications, 14
　irregular, 143, 171
　from penetrating keratoplasty, determinations of, 55, 63
　photorefractive keratectomy, 171
　with-the-rule, 55

Avellino dystrophy
　clinical features, 85
　discovery of, 85
　hereditary factors, 85
　pathology, 85
　phototherapeutic keratectomy treatment
　　clinical outcome studies, 104t
　　description of, 87
Axial distance, 53

Back scattering
　description of, 25
　illustration of, 26
Band keratopathy
　clinical outcome studies, 104t
　description of, 28
　masking fluids, 123
Basement membrane
　description of, 33
　epithelial basement membrane dystrophy
　　clinical features, 75–76, 76
　　histopathology, 76
　　pathogenesis, 76
　　pathology, 74–75
　　phototherapeutic keratectomy treatment, 76, 76
　post-phototherapeutic keratectomy, 37
BCVA, (see Best spectacle corrected visual acuity)
Best spectacle corrected visual acuity
　clinical studies using Summit Technology excimer lasers, 194, 194t–195t
　photorefractive keratectomy, 160t–164t, 167
Biber-Haab-Dimer dystrophy, (see Lattice dystrophy)
Bowman's layer
　anatomy of, 21
　description of, 21
　dystrophies
　　anterior membrane dystrophy of Grayson-Wilbrandt, 77–78
　　honeycomb dystrophy of Thiel and Behnke, 78
　　local anterior mucopolysaccharide accumulation, 78–79
　　Reis-Bücklers' dystrophy, (see Reis-Bücklers' dystrophy)
　　subepithelial mucinous corneal dystrophy, 78
　function of, 21
　phototherapeutic keratectomy effects, 39–40
　remnants after excimer laser, 11
　wound healing, 36–37

Carcinogenesis, wavelengths involved in, 10
Cataractogenesis, 10
Central crystalline dystrophy of Schnyder
　clinical features, 88
　illustration of, 89
　pathology, 88–90
　phototherapeutic keratectomy treatment
　　clinical outcome studies, 104t
　　description of, 90
Central islands
　characteristics of, 170
　definition of, 170
　from photorefractive keratectomy
　　description of, 170, 175, 178
　　incidence, 178
　　management, 179–180
　　pathogenesis, 178
　　phototherapeutic keratectomy retreatment, 179
　　risk factors, 178–179
　　subepithelial haze associated with, 178
Chondroitin sulfate, role in wound healing, 34t
Clinical studies, of excimer lasers
　Summit Technology
　　phototherapeutic keratectomy
　　　best corrected visual acuity, 194, 194t–195t
　　　patient survey results, 199t
　　　reepithelialization, 193, 194t
　　　refractive changes, 196t
　　　success rates, 196–197, 197t, 198
　　　techniques, 191, 193
　　　treatment categories, 191, 193, 193t
　　principal investigators, 192t–193t
　　refractive errors
　　　high-myopia, 165–166
　　　low-myopia, 158–159, 160t–161t
　VISX
　　phototherapeutic keratectomy

Clinical studies, of excimer lasers (contd.)
    complications, 204, 204t
    intraoperative technique, 202–203
    laser calibration, 202, 203t
    patient selection criteria, 201, 202t
    perioperative preparation, 201–202
    postoperative management, 203
    preoperative examination, 201
    results, 203t–204t, 203–204
    refractive errors
        high-myopia, 166
        low-myopia, 158–159, 160t–161t
Cogan's microcystic dystrophy, (see Epithelial basement membrane dystrophy)
Collagen fibrils, in stroma
    degradation, regulation of, 41
    description of, 22
    effect on transparency, 24
    light scattering and, 28
    resorption of, 41
Complications
    photorefractive keratectomy
        astigmatism, 171
        central islands
            description of, 170, 175, 178
            pathogenesis of, 178
            phototherapeutic keratectomy retreatment, 179
            risk factors, 178–179
            subepithelial haze associated with, 178
            treatment options, 179–180
        corneal haze, 167–168
        corneal topographic mapping, for phototherapeutic keratectomy retreatment, 175, 176–177
        decentration
            description of, 169–170
            phototherapeutic keratectomy retreatment, 183, 184–185, 185–186
        glare, 168–169
        haloes, 168–169
        ocular hypertension, 168
        subepithelial haze, phototherapeutic keratectomy retreatment
            classifications, 180–181
            description of, 180–181
            risk factors, 181–182
        surface irregularities, phototherapeutic keratectomy retreatment, 175
        undercorrections, phototherapeutic keratectomy retreatment, 180
    phototherapeutic keratectomy
        decentration, 147
        endothelial cell damage, 148
        epithelial defects
            persistent defects, 150
            recurrent erosions, 150
        graft rejection, 150–151
        herpes simplex virus reactivation, 150
        hyperopia, 15, 118, 144–145
        incomplete treatment, 147–148
        pain, 148–150
        refractive changes
            hyperopic shift, 144–146
            irregular astigmatism, 143–144
            myopic shift, 146
        subepithelial scarring, 151, 151
Cornea
    anatomy of
        Descement's layer, 22
        endothelium, 22–23
        epithelium, (see Epithelium)
        overview, 21
        stroma, 21–22
    curvature
        radius of curvature, 53
        videokeratoscope determinations, 53–54
    dystrophies, (see Corneal dystrophies)
    flattening
        correction of, 114
        hyperopia associated with, 118
    haze, (see Haze)
    innervation, 148–149
    surface irregularities, (see Surface irregularities)
    topography
        applications
            astigmatism determinations, 55–56
            keratoconus, 56–57
            post-photorefractive keratectomy, 175, 176–177
            post-phototherapeutic keratectomy conditions
                central island, 60–61
                hyperopic shift, 58–61
                surface irregularities after wound healing, 55, 61–62
                treatment decentration, 57
        for decentration evaluations, 184–185
        historical attempts to determine, 51
        methods to measure
            keratoscope, 51
            photokeratoscope, 51–52
            principles of, 52–53
            radius of curvature, 53
    transparency
        definition of, 23–24
        edema effects, 24
        endothelium functions, 23
        keratan sulfate effects, 34t
        scattering, 23–24
        stroma effects, 40
Cornea verticillata of Fleischer, (see Vortex dystrophy)
Corneal dystrophies, phototherapeutic keratectomy of
    band keratopathy, 28
    Bowman's layer dystrophies
        anterior membrane dystrophy of Grayson-Wilbrandt, 77–78
        honeycomb dystrophy of Thiel and Behnke, 78
        local anterior mucopolysaccharide accumulation, 78–79
        Reis-Bücklers' dystrophy, (see Reis-Bücklers' dystrophy)
        subepithelial mucinous corneal dystrophy, 78
    central crystalline dystrophy of Schnyder
        clinical features, 88
        illustration of, 89
        pathology, 88–90
        phototherapeutic keratectomy treatment, 90
    clinical outcome, (see also specific dystrophy)
        functional improvement, 103
        studies to determine, 104t
    complications, 114–115
    corneal erosions, (see Recurrent erosion syndrome)
    corneal surface irregularities, 64
    epithelial dystrophies
        anterior crocodile mosaic shagreen of Vogt, 74
        epithelial basement membrane dystrophy
            clinical features, 75–76, 76
            histopathology, 76
            pathogenesis, 76
            pathology, 74–75
            phototherapeutic keratectomy treatment, 76, 76
        juvenile hereditary epithelial dystrophy
            description of, 73
            pathology, 73–74
            phototherapeutic keratectomy treatment, 74
        vortex, 74
    gelatinous drop-like dystrophy
        clinical features, 90
        pathology, 90
        phototherapeutic keratectomy treatment, 91
    glare, 28
    recurrent erosion syndrome, (see Recurrent erosion syndrome)
    side effects, 114
    stromal dystrophies
        Avellino
            clinical features, 85
            discovery of, 85
            hereditary factors, 85
            pathology, 85

Index

phototherapeutic keratectomy treatment
    clinical outcome studies, 104t
    description of, 87
  discovery of, 79
  granular
    clinical features, 79, 79–80
    differential diagnosis, 80
    pathology, 80
    phototherapeutic keratectomy treatment
      clinical outcome, 104t, 107t, 108
      description of, 80–81
      illustration of, 80–81
  lattice
    familial predilection, 82
    illustration of, 81
    phototherapeutic keratectomy treatment
      clinical outcome, 13, 104t, 107t
      description of, 85
      illustration of, 86
    type I
      clinical features, 81, 83
      illustration of, 84
      pathology, 84–85
    type II
      clinical features, 83–84
      discovery of, 83
      pathology, 84–85
      vestibulocochleopathy and, relationship between, 83
  macular
    clinical features, 87
    pathology, 87–88
    phototherapeutic keratectomy treatment
      clinical results, 113
      description of, 88
  surgical technique
    epithelial removal
      laser ablation, 100
      manual debridement, 100
      overview, 99–100
    postoperative management, 101, 103
    preoperative preparation, 99
    stromal ablation, 100–101, 101–102
Corneal light reflex, 51–52
Corticosteroids
  complications from long-term use, 103
  corneal haze, 167–168, 179, 182
  post-phototherapeutic keratectomy, 43, 101, 103, 149
  topical, 149
Curvature, of cornea
  radius of curvature, 53
  videokeratoscope determinations, 53–54

Debridement, of epithelium
  manual, 100
  mechanical
    elevated corneal nodules, 118, 121
    modifications, 121
Decentration
  corneal topographic mapping to evaluate, 184–185
  description of, 147
  management of, 147
  photorefractive keratectomy
    corneal topography mapping, 184–185
    description of, 169–170
    patterns, 185
    phototherapeutic keratectomy treatment, 185–186
Decorin, (see Dermatan sulfate)
Defects, (see specific defect)
Dermatan sulfate
  description of, 34
  haze associated with, 41
  wound healing and, 34t, 41
Descemet's membrane
  composition of, 22
  description of, 21
  excimer laser ablation effects, 11
  function of, 22
  post-phototherapeutic keratectomy effects, 41–42
Diclofenac sodium, 101, 103, 149, 182
Dioptric power
  definition of, 53
  videokeratoscope determinations, 53–54
Distortion plots, 52
Dystrophic recurrent erosion, (see Epithelial basement membrane dystrophy)
Dystrophy, (see specific dystrophy)

ECM, (see Extracellular matrix)
Edema, corneal transparency and, 24
EGF, (see Epidermal growth factor)
Elevated lesions, (see Nodules, corneal)
Emmetropia, 166
Endothelium, corneal
  anatomy of, 22
  functions of, 22–23
  phototherapeutic keratectomy effects
    cell damage, 148
    description of, 41–42
  pump-leak hypothesis, 23
Epidermal growth factor
  description of, 35
  role in wound healing after phototherapeutic keratectomy, 42, 44
Epikeratophakia, powered lenticules for, 14
Epithelial basement membrane dystrophy
  clinical features, 75–76, 76
  histopathology, 76
  pathogenesis, 76
  pathology, 74–75
  phototherapeutic keratectomy treatment, 76, 76
Epithelial cells, role in corneal wound healing
  general injury
    interactions with extracellular matrix, 35
    migration to wound, 36
  post-phototherapeutic keratectomy, 38
Epithelium
  adhesion complex, 21
  anatomy of, 21
  cells, (see Epithelial cells)
  dystrophies
    anterior crocodile mosaic shagreen of Vogt, 74
    epithelial basement membrane dystrophy
      clinical features, 75–76, 76
      histopathology, 76
      pathogenesis, 76
      pathology, 74–75
      phototherapeutic keratectomy treatment
        clinical outcome studies, 104t
        description of, 76
        illustration of, 76
    juvenile hereditary epithelial dystrophy
      description of, 73
      pathology, 73–74
      phototherapeutic keratectomy treatment, 74
    phototherapeutic keratectomy technique
      clinical outcome studies, 104t
      laser ablation, 99–100
      manual debridement, 100
      preoperative preparation, 99
    vortex, 74
  growth factor effects, 35
  illustration of, 34
  layers of, 33
  persistent defects, 150
  as surface modulator, 120
  wound healing, 37–38, 38
Excimer lasers
  ablative decomposition
    calibrations, 10
    depth, 9–10
    mechanism of action, 8–9
    threshold, 9
  aperture wheels, 8, 8–9
  beam intensity, 7
  clinical studies, (see Clinical studies)
  components of, 6–7
  description of, 6
  erbium:yttrium-aluminum-garnet, 5
  gas mixtures, 6

Excimer lasers (contd.)
  neodymium, 5
  patient considerations, 8
  photorefractive keratectomy, (see Photorefractive keratectomy)
  phototherapeutic keratectomy, (see Phototherapeutic keratectomy)
  refractive surgical applications
    astigmatism, 14
    hyperopia, 14
    myopia, 14
    plano or powered lenticules for epikeratophakia, 14
  safety issues
    carcinogenesis, 10
    cataractogenesis, 10
    corneal integrity, 11
    heat generation, 10–11
    mutagenesis, 10
  Summit Technology, (see Summit Technology excimer lasers)
  VISX, (see VISX excimer laser)
Extracellular matrix, of corneal stroma
  cell-matrix interactions, 35
  composition, 33
  fibronectin, 34–35
  integrins, 35
  laminin, 34
  proteoglycans, 33–34
  transforming growth factor effects, 42

Familial amyloid polyneuropathy, (see Lattice dystrophy, type II)
FGF, (see Fibroblast growth factor)
Fibrils, collagen
  degradation, regulation of, 41
  description of, 22
  effect on transparency, 24
  light scattering and, 28
  resorption of, 41
Fibroblast growth factor, 35
  role in corneal wound healing, post-phototherapeutic keratectomy, 43
  types of, 43
Fibronectin, role in corneal wound healing
  general injury, 34–35, 37
  post-phototherapeutic keratectomy, 37–38
Fingerprint/map/dot dystrophy, (see Epithelial basement membrane dystrophy)
Flattening, of central cornea
  correction of, 114
  hyperopia associated with, 118
Fluorometholone, 149
Flurbiprofen sodium, 149
Forward scattering
  clinical studies, 27
  description of, 25
  glare associated with, 25
  illustration of, 26–27
  soft contact lens vs. hard contact lens, 26–27

Gas mixtures, for excimer laser
  argon fluoride, (see Argon fluoride gas)
  types, 6
Gelatinous drop-like dystrophy
  clinical features, 90
  pathology, 90
  phototherapeutic keratectomy treatment
    clinical outcome studies, 104t
    description of, 91
Glare
  from forward scattering, 25–27
  phototherapeutic keratectomy treatment, 28
  post-photorefractive keratectomy, 168–169
Glycosaminoglycans
  in macular dystrophy, 87
  role in corneal transparency, 41
Graft rejection, corneal transplant failure and, 150–151
Granular dystrophy
  clinical features, 79, 79–80
  differential diagnosis, 80
  pathology, 80
  phototherapeutic keratectomy treatment
    clinical outcome, 12, 104t, 107t, 108
    description of, 80–81
    illustration of, 80–81, 108
    for recurrent dystrophy, 109
    stromal ablation, 100
Grayson-Wilbrandt dystrophy, (see Anterior membrane dystrophy of Grayson-Wilbrandt)
Growth factors, (see also specific growth factor)
  description of, 33
  role in corneal wound healing, 35, 42–43
  types of, 35

HA, (see Hyaluronic acid)
Halo
  postexcimer complication, 25
  post-photorefractive keratectomy, 168–169
Haze
  ablation zone size and, 168
  associated symptoms, 25
  classification of
    post-phototherapeutic keratectomy, 103
    using slit lamp examination, 11
  corneal topography, (see Cornea, topography)
  density increases in, factors that affect, 13
  illustration of, 41
  light scattering
    back scattering, 25, 26
    description of, 24–25
    forward scattering, 25–27, 26–27
  management of
    options, 168
    retreatment, 179
    steroid treatment
      clinical studies, 25, 43–44
      description of, 13, 43–44
  post-photorefractive keratectomy
    classifications, 181
    description of, 167–168, 180–181
    medical management, 182
    risk factors, 181–182
    surgical management, 182–183
  post-phototherapeutic keratectomy
    extracellular matrix effects, 41
    therapeutic approaches, 43–44
  resolution of, 25
  slit lamp evaluations, 11, 12–13
  steroid treatment, 13
Healing, (see Wound healing)
Heat generation, 10–11
Hemidesmosomes, 33
Heparan sulfate, 34, 34t
Herpes simplex virus reactivation, from photorefractive keratectomy, 150
Honeycomb dystrophy of Thiel and Behnke, 78
HSV reactivation, (see Herpes simplex virus reactivation)
Hyaluronic acid, 40
Hypermetropia, 14
Hyperopia
  ablation depth and, correlation between, 146
  from corneal flattening, 118, 144–145
  corneal topographic mapping of, 58–61
  excimer laser treatment
    aperture wheels, 9
    applications, 14
  post-phototherapeutic keratectomy, 144–146, 145

Inflammation mediators, 149
Integrins
  description of, 35
  receptor function of, 35
  role in corneal wound healing, 33, 35, 36t
  types of, 36t
Irregularities, (see Surface irregularities)
Islands, (see Central islands)
Isodioptric display, 52

Juvenile hereditary epithelial dystrophy
  description of, 73
  pathology, 73–74

Index

phototherapeutic keratectomy treatment
  clinical outcome studies, 104t
  description of, 74

Keratan sulfate
  macular dystrophy and, 87
  role in wound healing, 34t
Keratoconus
  apical scars in, phototherapeutic keratectomy treatment, clinical outcome studies, 104t
  corneal topography mapping for, 55–56
Keratocytes
  description of, 21–22
  role in corneal wound healing, 37
Keratoscopes
  development of, 51
  photokeratoscope, 51–52
  videokeratoscope, (see Videokeratoscope)
Ketorolac tromethamine, 149, 182
KrF, (see Krypton fluoride)
Krypton fluoride, 3

Lakes, 24
Lamellar keratoplasty, for corneal nodules and scars
  description of, 117
  phototherapeutic keratectomy and, comparison, 117–118
Laminins, 34
Laser assisted in situ keratomileusis, 179
Laser thermokeratoplasty, 4
Lasers, excimer, (see Excimer lasers)
LASIK, (see Laser assisted in situ keratomileusis)
Lattice dystrophy
  familial predilection, 82
  illustration of, 81
  phototherapeutic keratectomy treatment
    clinical outcome, 13, 104t, 107t
    description of, 85
    illustration of, 86
    for recurrent dystrophy, 110–111
  type I
    clinical features, 81, 83
    illustration of, 84
    pathology, 84–85
  type II
    clinical features, 83–84
    discovery of, 83
    pathology, 84–85
    vestibulocochleopathy and, relationship between, 83
Leukocytes, 37
Leukotrienes, 149
Light scattering
  after phototherapeutic keratectomy, 40–41, 41

back scattering
  description of, 25
  illustration of, 26
  corneal defect configuration effects, 25
  description of, 24–25
  forward scattering
    clinical studies, 27
    description of, 25
    glare associated with, 25
    illustration of, 26–27
    soft contact lens vs. hard contact lens, 26–27
Line of sight
  corneal light reflex and, 52
  definition of, 52
Local anterior mucopolysaccharide accumulation, 78–79

Macular dystrophy
  clinical features, 87
  pathology, 87–88
  phototherapeutic keratectomy treatment
    clinical results, 113
    description of, 88
Masking fluids, (see also Surface modulators)
  application of, 123
  function of, 119, 144
  for laser epithelial ablation, 100
  properties of, 119, 120
  types of, 119, 121
Mechanical debridement, for epithelial removal
  description of, 100
  elevated corneal nodules, 118, 121
  modifications, 121
Meesmann's dystrophy, (see Juvenile hereditary epithelial dystrophy)
Metalloproteinases, 41
Mitomycin C, 43
MPS, (see Mucopolysaccharides)
Mucopolysaccharides, macular dystrophy and, 87–88
Mutagenesis, 10
Myopia
  corneal flattening, topographic mapping to determine, 55, 63, 64
  laser refractive keratectomy, 14
  photorefractive keratectomy results
    high-myopia, 162t–163t, 164–166
    low- and high-myopia, 166
    low-myopia, 158–159, 160t–161t
  post-phototherapeutic keratectomy, 146
  refractive errors, treatment of, 101
  and subepithelial scarring and haze, correlation between, 14

Neodymium lasers, 5
Nodules, corneal
  phototherapeutic keratectomy

advantages of, 117–118
challenges associated with, 121
difficulties associated with, 118–119
elevated lesions
  central
    description of, 118
    treatment approaches, 125, 127
  large types, surgical approach, 123
  manual debridement, 118
  multiple
    description of, 127
    treatment algorithm, 128
  therapeutic approaches
    ablation, 124, 125
    mechanical debridement, 121
    surface modulators, 121, 123
lamellar keratoplasty, 117
laser pulse repetition rate settings, 119–120, 125
postoperative management, 127, 130
therapeutic approaches
  ablation, 124, 125
  mechanical debridement, 121
  treatment algorithm, 121
surface irregularities associated with, 118
surface modulators, 119–121
surgical technique, 124
treatment algorithm, 121
Nonsteroidal anti-inflammatory drugs (NSAIDs)
  corneal haze, 182
  phototherapeutic keratectomy, 69t, 149

Ocular hypertension, from photorefractive keratectomy, 168
Opacities
  from dystrophies, (see Corneal dystrophies)
  haze, (see Haze)
  nodules, (see Nodules)
  phototherapeutic keratectomy treatment, 99
  scars, (see Scars)
  in stroma
    classifications, 11
    illustration of, 67
    phototherapeutic keratectomy goals, 66–67
  surface irregularities, (see Surface irregularities)

PDGF, (see Platelet-derived growth factor)
Penetrating keratoplasty, 115
  astigmatism associated with, corneal topographic mapping to determine, 55, 63
  clinical applications
    central crystalline dystrophy of Schnyder, 89

Penetrating keratoplasty (contd.)
    herpetic corneal scars, 150
    lattice dystrophy, 85
    disadvantages of, 99
Perlecan, (see Heparan sulfate)
Photoablation, (see also Ablation)
    biological mediators, 149
    description of, 6
    illustration of, 4
    phototherapeutic keratectomy, (see Phototherapeutic keratectomy)
Photocoagulation, 3, 4, 5
Photodisruption, 4, 5
Photokeratoscope, 51
Photorefractive keratectomy
    clinical studies and outcomes
        best spectacle corrected visual acuity, 160t–164t, 167
        high-myopia, 162t–163t, 164–166
        low- and high-myopia, 166
        low-myopia, 158–159, 160t–161t
        patient selection, 157
        postoperative management, 158
        preoperative evaluation, 158
        Summit Technology UV200, 159t
        surgical technique, 158
        uncorrected visual acuity, 166–167
        VISX, 159t
    complications
        astigmatism, 171
        central islands
            description of, 170, 175, 178
            pathogenesis of, 178
            phototherapeutic keratectomy retreatment, 179
            risk factors, 178–179
            subepithelial haze associated with, 178
            treatment options, 179–180
        corneal haze, 167–168
        corneal topographic mapping for phototherapeutic keratectomy retreatment, 175, 176–177
        decentration
            description of, 169–170
            phototherapeutic keratectomy retreatment, 183, 184–185, 185–186
        glare, 168–169
        haloes, 168–169
        ocular hypertension, 168
        subepithelial haze, phototherapeutic keratectomy retreatment
            classifications, 180–181
            description of, 180–181
            risk factors, 181–182
        surface irregularities, phototherapeutic keratectomy retreatment, 175
        undercorrections, phototherapeutic keratectomy retreatment, 180
    concomitant therapy with phototherapeutic keratectomy, 101
    development of, 157
    healing patterns, 181
    hyperopia treatment, 14
Phototherapeutic keratectomy
    advantages of, versus conventional therapies, 197–198, 199t
    clinical applications
        corneal dystrophies, (see Corneal dystrophies)
        glare, (see Glare)
        haze, (see Haze)
        nodules, (see Nodules, corneal)
        recurrent erosion syndrome, (see Recurrent erosion syndrome)
        refractive errors, (see Refractive errors)
        scars, (see Scars, corneal)
        surface irregularities, (see Surface irregularities)
        wound healing, (see Wound healing)
    clinical studies, (see Clinical studies)
    complications, (see Complications)
    contraindications, 67, 68
    description of, 14–15
    development of, 14
    intrastromal, 5
    lasers for
        Summit Technology, (see Summit Technology excimer laser)
        VISX, (see VISX excimer laser)
    and photorefractive keratectomy, (see Photorefractive keratectomy)
    postoperative regimen, (see Postoperative regimen)
    preoperative considerations, (see Preoperative considerations)
Photovaporization, 4, 5
Plácido's disk
    description of, 51
    development of, 51
    keratoscopes, (see Keratoscopes)
Platelet-derived growth factor, 35, 43
Point and shoot technique, 193
Polymethyl methacrylate (PMMA), 6, 119
Polymorphonuclear neutrophils (PMN), 37
Postoperative regimen, for phototherapeutic keratectomy
    examination, 69–70
    instructions to patient, 69
    medications, 69, 69t
PPK, (see Prismatic photokeratectomy)
Preoperative considerations, for phototherapeutic keratectomy
    corneal topography, (see Cornea, topography)
    patient criteria, 65
    recommended examinations, 66t
    slit lamp examination, 65
    surgical preparation
        anesthesia, 66
        based on pathology, 66, 67
        elevated vs. smooth opacities, 66–67, 67–68
        laser setup, 66
        patient positioning, 66–67
Primary familial amyloidosis of the cornea, (see Gelatinous drop-like dystrophy)
Prismatic photokeratectomy, indications, 15
PRK, (see Photorefractive keratectomy)
Prostaglandins, 149
Proteoglycans, (see also specific proteoglycan)
    description of, 33–34
    formation of, 33
    post-phototherapeutic keratectomy response
        description of, 37
        haze, 41
    types of, 34t
PTK, (see Phototherapeutic keratectomy)
Pump-leak hypothesis, 23

Radial keratotomy, 134
Radius of curvature, 53
Recurrent erosion syndrome
    conventional therapy, 99, 133
    discovery of, 133
    manual debridement of epithelium, 100
    pathogenesis of, 133
    phototherapeutic keratectomy
        adhesion structure reassembly, 136t–137t
        clinical outcome
            results, 133–134, 134t–135t, 138
            studies, 104t–105t, 193
        complications, 133
        contraindications, 139–140
        histopathological findings, 134–136, 137–138
        incidence, 150
        operative techniques, 136–137
        principles, 133
        for recalcitrant recurrent erosion syndrome, 139–140
Reepithelialization, (see also Epithelium)
    duration of, 11
    post-phototherapeutic keratectomy, 37, 38

Index

Refractive errors
  astigmatism, 14
  hyperopia, (see Hyperopia)
  myopia, (see Myopia)
  from phototherapeutic keratectomy
    hyperopic shift, 144–146
    irregular astigmatism, 143–144
    myopic shift, 146
    plano or powered lenticules for epikeratophakia, 14
Refractive power, 53
Reis-Bücklers' dystrophy
  clinical features, 77
  discovery of, 76–77
  etiology, 77
  manual epithelial debridement, 100
  pathology, 77
  phototherapeutic keratectomy treatment
    clinical outcome studies, 104t, 106t
    description of, 67, 77
    illustration of, 68
RES, (see Recurrent erosion syndrome)
RK, (see Radial keratotomy)

Salzmann's nodular degeneration
  ablation strategies, 124, 125
  clinical outcome
    results, 129–130
    studies, 104t
  multiple nodules, 127, 129–130
  surgical goals, 67–68
Scars, corneal
  conventional surgical treatment methods, 99
  herpetic, 150
  phototherapeutic keratectomy
    advantages of, 117–118
    challenges associated with, 121
    clinical outcome
      results, 113–114, 122, 126–128
      studies, 104t
    corneal topographic mapping, 123
    difficulties associated with, 118–119
    elevated lesions
      central
        description of, 118
        treatment approaches, 125, 127
      large types, surgical approach, 123
      manual debridement, 118
      multiple, 127
      therapeutic approaches
        ablation, 124, 125
        mechanical debridement, 121
        surface modulators, 121, 123
        treatment algorithm, 121
    lamellar keratoplasty and, comparison, 117–118
    laser pulse repetition rate settings, 119–120, 125
  postoperative management, 127, 130
  stromal scars, 147–148
  surgical goals, 67, 68
  subepithelial, 151, 151
Scattering
  definition of, 23–24
  light
    after phototherapeutic keratectomy, 40–41, 41
    back scattering
      description of, 25
      illustration of, 26
      corneal defect configuration effects, 25
    description of, 24–25
    forward scattering
      clinical studies, 27
      description of, 25
      glare associated with, 25
      illustration of, 26–27
      soft contact lens vs. hard contact lens, 26–27
Schnyder's dystrophy, (see Central crystalline dystrophy of Schnyder)
Slit lamp evaluations
  for haze
    classifications, 11
    illustrations, 12–13
  for phototherapeutic keratectomy, 65
Steroids, (see Corticosteroids)
Stroma
  ablation, 100–101
  anatomy of, 21–22, 33
  composition, 24
  dystrophies
    Avellino
      clinical features, 85
      discovery of, 85
      hereditary factors, 85
      pathology, 85
      phototherapeutic keratectomy treatment, 87
    discovery of, 79
    granular
      clinical features, 79, 79–80
      differential diagnosis, 80
      pathology, 80
      phototherapeutic keratectomy treatment
        clinical outcome, 104t, 107t, 108
        description of, 80–81
        illustration of, 80–81
    lattice
      familial predilection, 82
      illustration of, 81
      type I
        clinical features, 81, 83
        illustration of, 84
        pathology, 84–85
        phototherapeutic keratectomy treatment, 85, 86
      type II
        clinical features, 83–84
        discovery of, 83
        pathology, 84–85
      type I, 85, 86
      vestibulocochleopathy and, relationship between, 83
    macular
      clinical features, 87
      pathology, 87–88
      phototherapeutic keratectomy treatment, 88
  extracellular matrix
    cell-matrix interactions, 35
    composition, 33
    fibronectin, 34–35
    integrins, 35
    laminin, 34
    proteoglycans, 33–34
    transforming growth factor effects, 42
  hydration, from endothelium, 23, 23
  inflammatory mediator release, 37
  lamellar, 22
  opacities
    haze, (see Haze)
    illustration of, 67
    phototherapeutic keratectomy goals, 66–67
  phototherapeutic keratectomy
    ablation methods, 100–101, 101–102
    postoperative healing, 40–41
  wound healing, 40–41
Subepithelial mucinous corneal dystrophy, 78
Summit Technology excimer lasers
  centration, 169
  clinical studies
    high-myopia, 165–166
    low-myopia, 158–159, 160t–161t
    phototherapeutic keratectomy
      best corrected visual acuity, 194, 194t–195t
      corneal dystrophies, 104t–107t
      patient survey results, 199t
      reepithelialization, 193, 194t
      refractive changes, 196t
      success rates, 196–197, 197t, 198
      techniques, 191, 193
      treatment categories, 191, 193, 193t
  principal investigators, 192t–193t
Surface irregularities
  after photorefractive keratectomy, 175
  corneal topographic mapping, 175
  phototherapeutic keratectomy treatment
    clinical outcome studies, 104t
    description of, 64

Surface modulators
  collagen gel, 123
  description of, 119
  epithelium, 120
  function of, 119
  masking fluids, (see Masking fluids)

Taper
  modified, 100–101, 125
  standard, 100, 125
Tenascin, role in corneal wound healing
  general injury, 37
  post-phototherapeutic keratectomy, 37
TGF-β, (see Transforming growth factor-β)
Thermokeratoplasty, 4, 5
Thromboxanes, 149
Topographic mapping of cornea, (see Cornea, topography)
Transforming growth factor
  α, 42
  β, 35
    corneal angiogenesis and, 42
    corneal wound healing and, 42, 44
    extracellular matrix effects, 42
    role in corneal wound healing, post-phototherapeutic keratectomy, 42
Transition zones, creation during stromal ablation, 100–101
Transparency, corneal
  definition of, 23–24
  edema effects, 24
  endothelium functions, 23
  keratan sulfate effects, 34t
  scattering, 23–24
  stroma effects, 40

UCVA, (see Uncorrected visual acuity)
Uncorrected visual acuity, post-photorefractive keratectomy, 166–167
Undercorrections, phototherapeutic keratectomy retreatment of, 180

Vertex, 52
Videokeratoscope
  algorithm designs, 54–55
  development of, 52
  principles of, 53
  spherical bias, 54
VISX excimer laser, clinical studies of
  high-myopia, 166
  low-myopia, 158–159, 160t–161t
  phototherapeutic keratectomy
    complications, 204, 204t
    corneal dystrophies, 104t–107t
    intraoperative technique, 202–203
    laser calibration, 202, 203t
    patient selection criteria, 201, 202t
    perioperative preparation, 201–202
    postoperative management, 203
    preoperative examination, 201
    results, 203t–204t, 203–204
Vortex dystrophy, 74

Wound healing
  approaches, 43–44
  Bowman's layer remnants, 11
  complications, 43
  of corneal anatomy
    Bowman's layer, 39–40
    Descemet's membrane, 41–42
    endothelium, 41–42
    epithelium, 37–39
    stroma, 40–41, 41
  corticosteroids, (see Corticosteroids)
  extracellular matrix effects
    cell-matrix interactions, 35
    composition, 33
    fibronectin, 34–35, 37
    growth factors, 35
    integrins, 35
    laminin, 34
    proteoglycans, 33–34
  growth factors
    epidermal growth factor, 42
    fibroblast growth factors, 43
    platelet-derived growth factor, 43
    transforming growth factor, 42
  keratocytes, 37
  overview, 33
  principles of, 35–37
  reepithelialization
    duration of, 11
    post-phototherapeutic keratectomy, 37, 38
    surface irregularities detected using corneal topographic mapping, 55, 61–62

Xenon bromide, 3, 6
Xenon chloride, 3, 6
Xenon fluoride, 3, 6